W9-BYX-937

MAKE it WORK!

TIME

Andrew Haslam

written by
David Glover

Photography: Jon Barnes
Consultant: Graham Peacock
Senior Lecturer in Science Education at
Sheffield Hallam University

WORLD BOOK / TWO-CAN

MAKE it WORK!
Other titles

Body
Building
Dinosaurs
Earth
Electricity
Flight
Insects
Machines
Photography
Plants
Ships
Sound
Space

First published in the United States in 1996 by
World Book Inc.
525 W. Monroe
20th Floor
Chicago
IL USA 60661
in association with Two-Can Publishing Ltd.

**For information on other World Book products,
call 1-800-255-1750, x 2238.**

ISBN: 0-7166-1730-7 (pbk.)
ISBN: 0-7166-1729-3 (hbk.)
LC: 96-60452

Printed in Hong Kong

1 2 3 4 5 6 7 8 9 10 99 98 97 96

Editor: Robert Sved
Designer: Lisa Nutt
Series Editor: Kate Graham
Managing Editor: Christine Morley
Managing Art Director: Carole Orbell
Production: Joya Bart-Plange
Additional photography: John Englefield
Series concept and design: Andrew Haslam and Wendy Baker

Thanks also to: Marisa Hawthorn, Lian Ng, James Perkins and everyone at Plough Studios

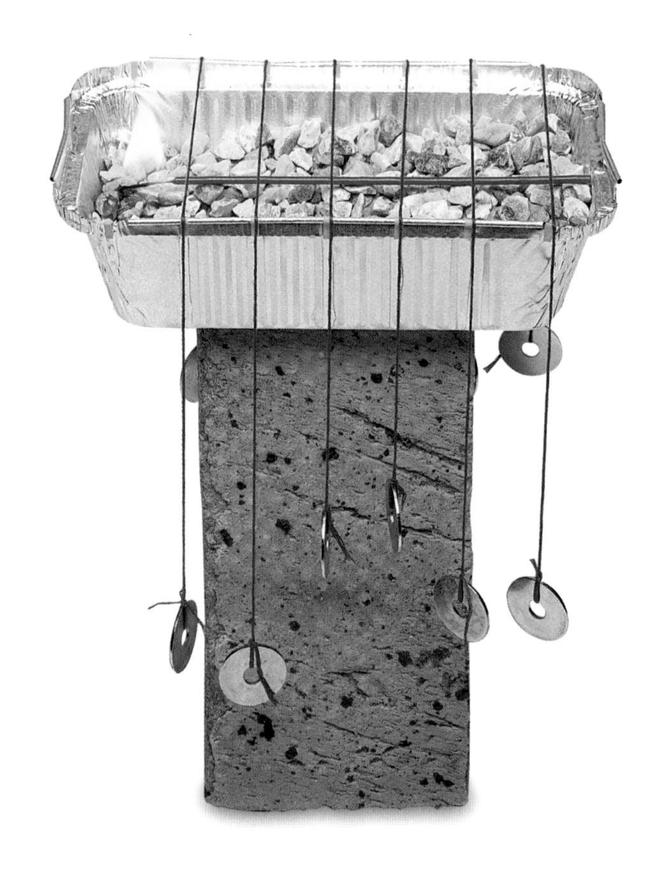

Contents

Words marked in **bold** in the text are explained in the glossary.

4 Making Time

Everyone needs to tell the time. We plan our lives with clocks and **calendars**. But do you know why there are 24 **hours** in a **day** or why every fourth **year** is a **leap year**? Can you tell the time by the stars?

MAKE it WORK!
In this book you will learn how our basic lengths of time—the day, the **month**, and the year—relate to the motion of Earth around the sun, and the moon around Earth. You will become a clockmaker yourself, investigate how different clocks work, and check to see if they keep good time. You will discover if time could ever run backward or if it is possible to travel through time.

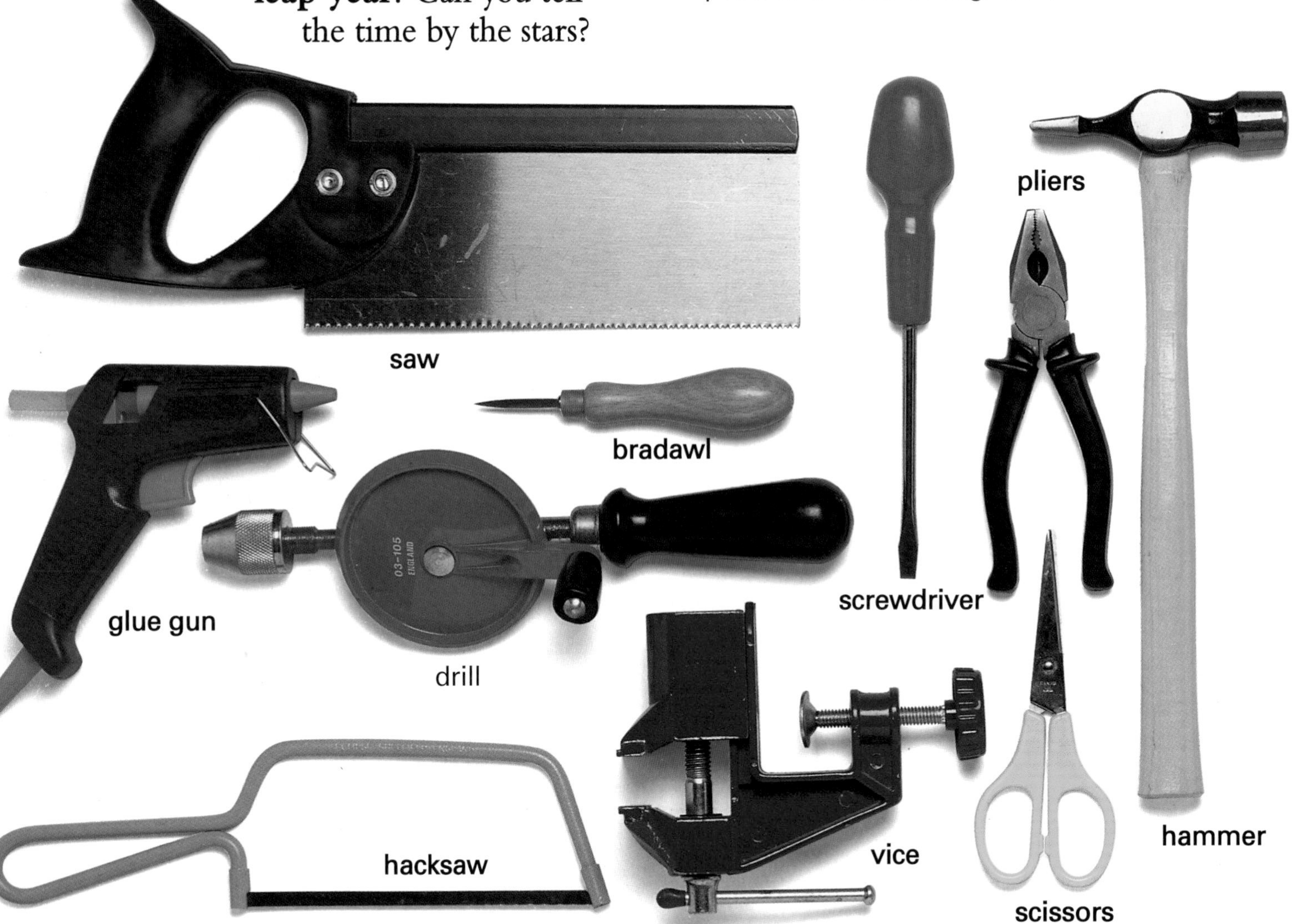

The story of time is an exciting mixture of science, history, and invention. Ancient stone circles, such as Stonehenge in England, were great shadow clocks that measured the passing seasons. Now science has given us **digital** watches that measure time in hundredths of a **second**.

You will need
You don't need expensive materials to explore time's mysteries. You can build most of the models from simple materials such as cardboard, wood, straws, plastic containers, and other odds and ends. You will need some tools to cut the materials, a drill to make holes, and glue, tape, and screws to put things together.

Pencils, a ruler, a protractor, a calculator, and a notebook will help you to plan your projects and to record your observations. Stencils are useful for making neat scales and labels on your models. You will need a magnetic **compass** to point your **sundials** in the right direction and a watch with a seconds display to check that your clocks work accurately. Pens and paints of different colors will make your clocks and models easier to read.

Drilling

To stop the drill bit from slipping, use a bradawl to start the holes. Then finish the holes off with a hand drill.

Using candles

Some projects use candles. When you light candles ask an adult to help you. Remember not to leave lit candles unattended at any time.

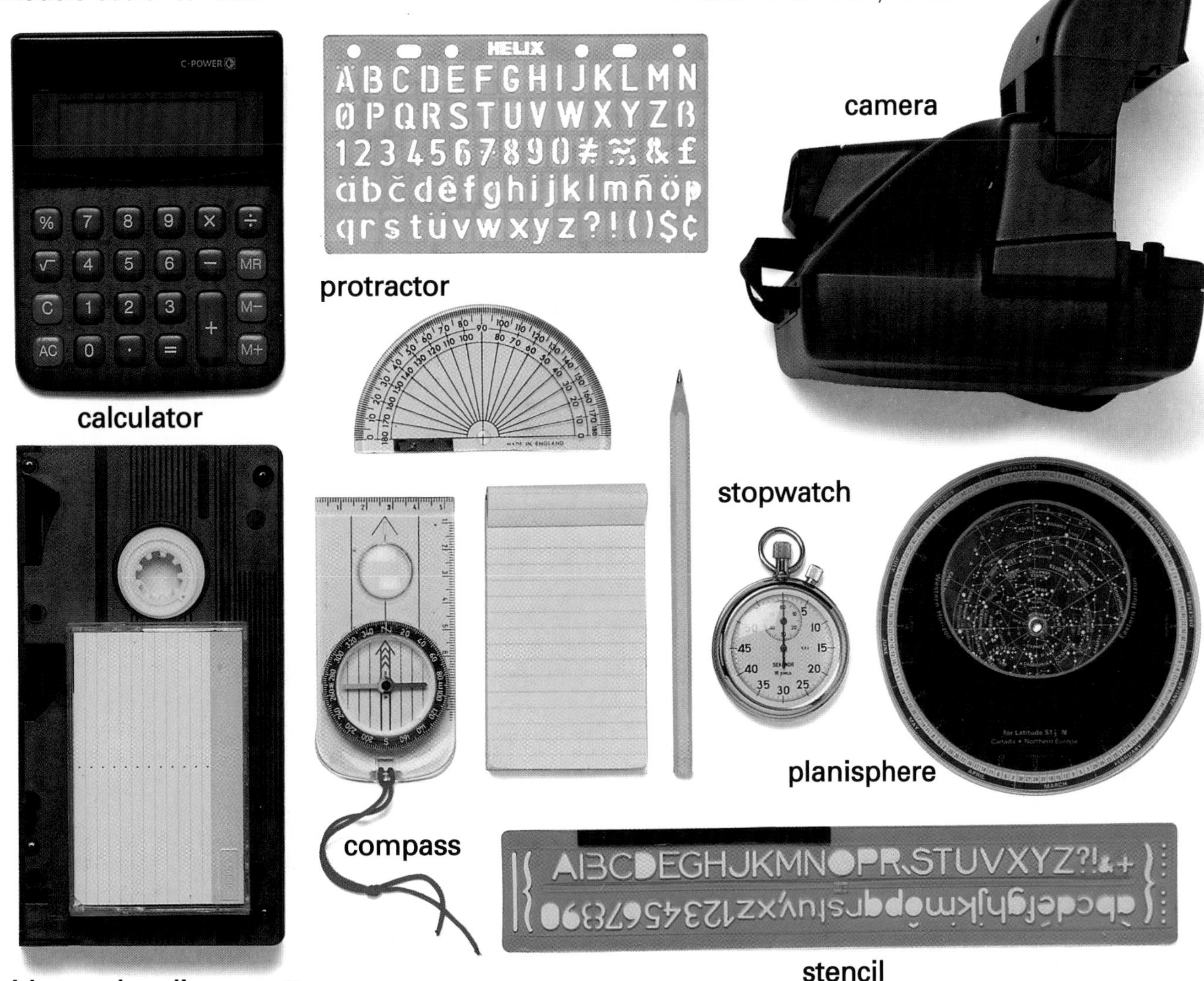

To measure time at night you will need a star chart or a **planisphere** to help you find the positions of stars in the sky. You can find charts in astronomy books or buy them separately. For two projects you will need to take pictures with an instant camera or a video camera. Ask permission before borrowing cameras from adults or friends.

Safety!

Sharp tools are dangerous! Always take care when you use them and ask an adult to help you. Make sure that anything you are cutting or drilling is held firmly so that it cannot slip. A small table vise is ideal for holding pieces of wood in place.

6 Night and Day

The old saying "As sure as the day follows night" means that something is certain to happen. The regular cycle of day and night, caused by Earth's spin, has marked out the passage of time since Earth was formed, more than 4.5 billion years ago. All living things on Earth, including ourselves, respond to this natural clock.

MAKE it WORK!
Earth makes one complete turn on its **axis** each day. When the place where you live is on the side of Earth that faces the sun, it is daylight. When Earth has turned so that your home is in the shadow, it is night.

A day is divided into 24 hours. Half of Earth is in the sun and half is in shade, so you might think that there would be 12 hours of daylight and 12 hours of darkness each day. This is not the case. In summer, there are more hours of daylight than darkness. In winter, night lasts longer than day. Making these models and this chart will help you understand why the hours of daylight and darkness change with the seasons.

You will need

- a plastic ball
- a small sponge-rubber ball
- a flashlight for the sun's light
- two short pieces of cardboard tube
- glue
- pens
- paints
- paper

1 Paint the smaller ball to make a model Earth. Paint the equator in red on your model. Paint the larger ball to make the sun.

2 Mount Earth and the sun on short lengths of cardboard tube. Earth's axis is tilted at 23°, so you should mount your model so that the equator is tilted, as shown at right.

3 In a dark room, shine a flashlight directly onto the side of Earth. You will see that half of the world is in darkness while the other is in light.

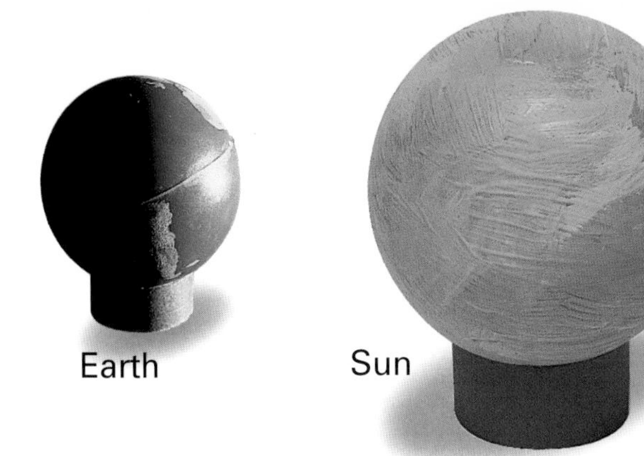

In December, Earth's Southern **Hemisphere** is tilted toward the sun, as above. As Earth spins, places south of the equator are on the sunlit side of Earth longer than on the dark side, so days are longer than nights.

As Earth spins, it moves around the sun. This gives us the seasons. After six months Earth is on the opposite side of the sun. Without turning Earth, place it on the other side of the sun and shine a flashlight onto Earth.

The model now shows the position of Earth in June. Now the Northern Hemisphere is tilted toward the sun, and days are longer than nights on this part of Earth. It is summer in the north and winter in the south.

Plotting day and night

Make a chart from colored paper, as shown below. The colored blocks along the top and bottom show the 12 months. The 24 hours of the day are marked up the side, and noon is the red line across the center.

Each week, look in a newspaper to find the times of sunrise and sunset. Mark these times on your chart and paint the hours of darkness blue. The yellow band across the middle of the chart shows how the hours of daylight vary through the year in the place where you live. The chart below is for Greenwich, England.

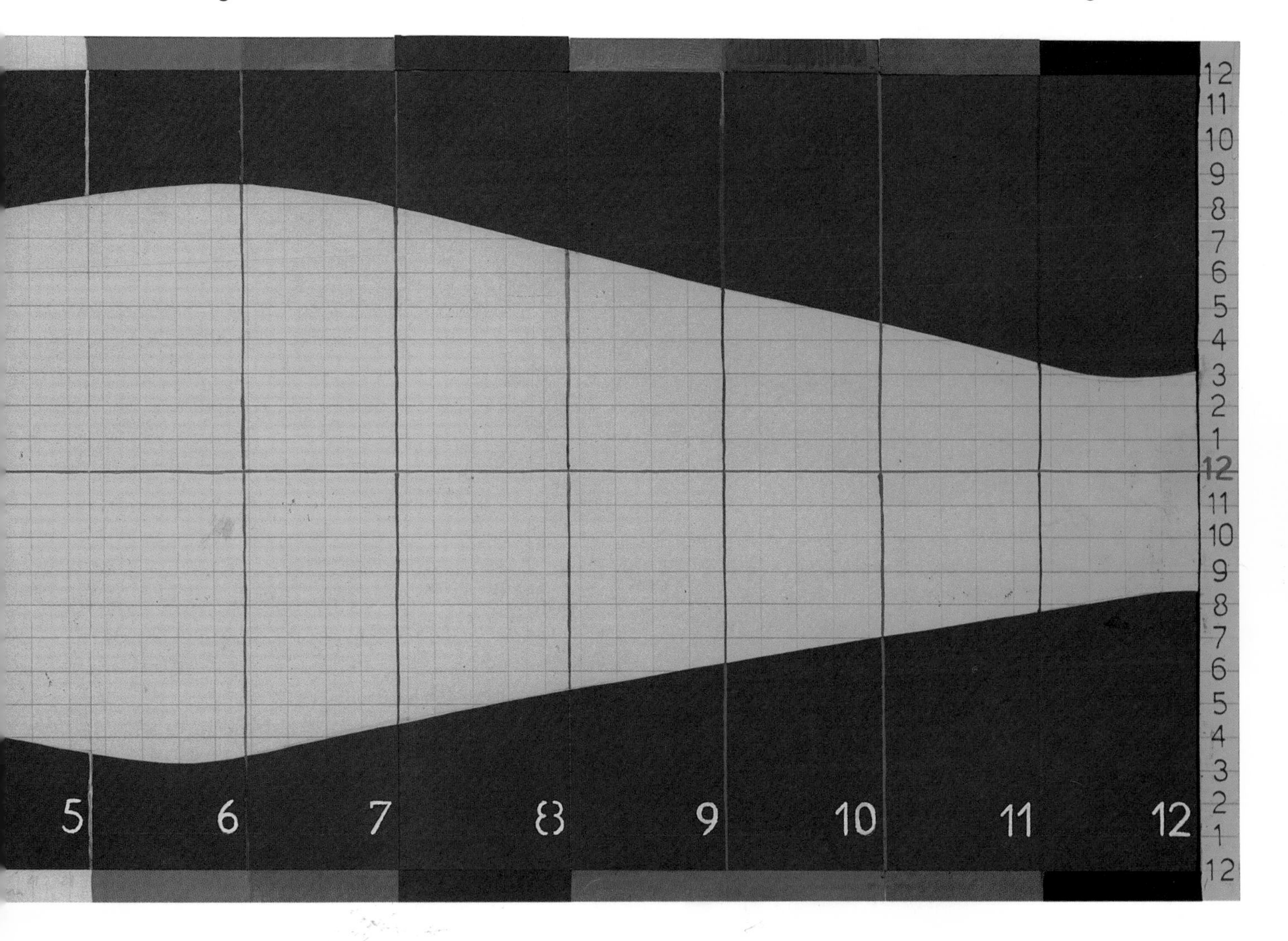

8 The Solar Year

As well as the natural cycle of day and night there are two more cycles that mark time for us. A year is the time it takes for Earth to circle the sun. A month is very nearly the time taken for the moon to circle Earth.

1 The three balls are the sun (largest ball), Earth and the moon (smallest ball). Paint them in realistic colors, as shown below.

2 Draw around a large dinner plate and the bases of cups or cans to make three cardboard circles of different sizes, as shown below. Cut out the circles with scissors, and paint them blue.

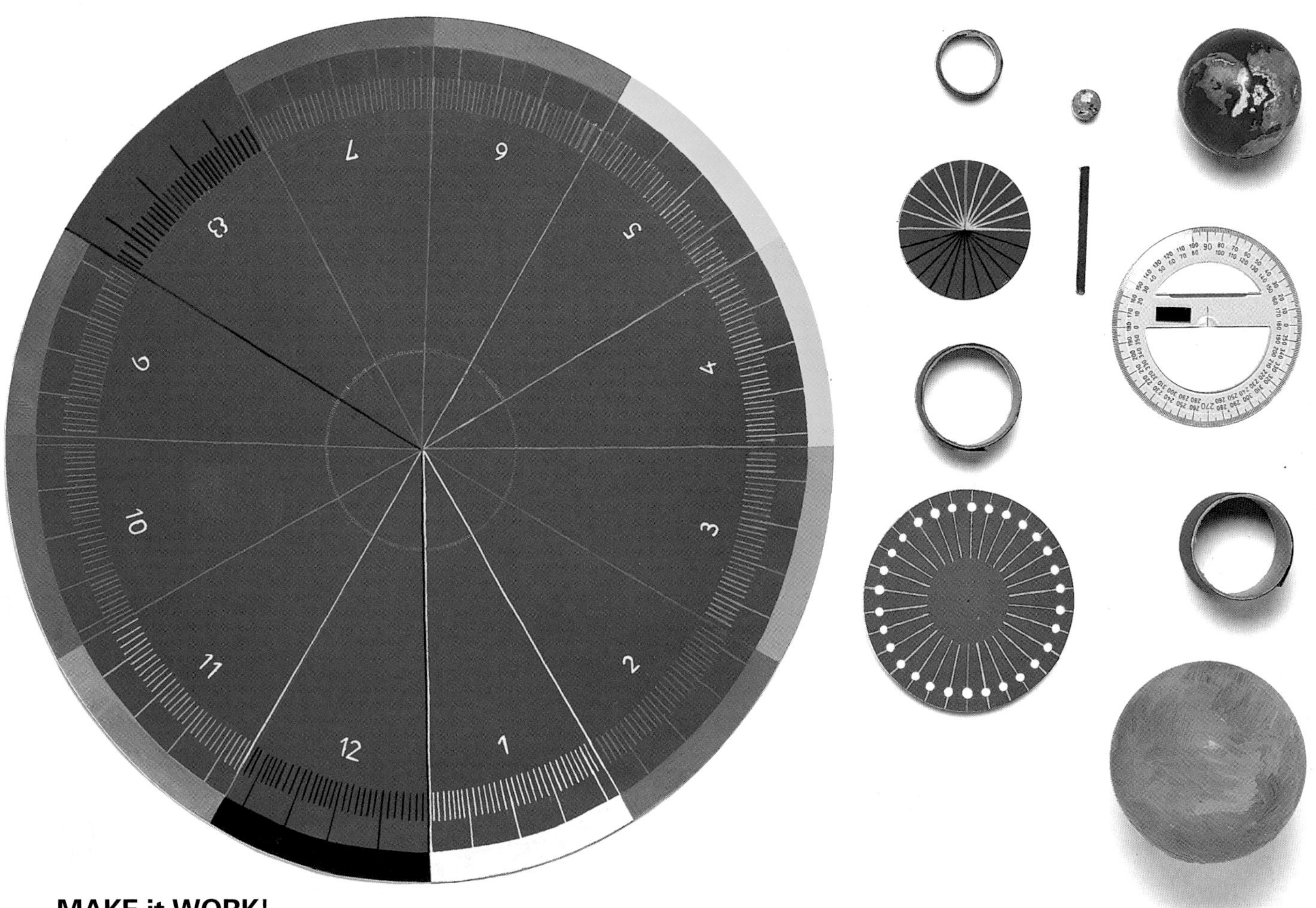

MAKE it WORK!
This **solar** calendar shows the motions of the sun, Earth, and the moon through the year. You can move Earth and the moon around in their **orbits** to follow the passage of the seasons as the days progress.

You will need

large, medium, and small rubber balls
short lengths of cardboard tube
4″ length of dowel
paints and brushes
scissors
a protractor
thick cardboard

3 You are now ready to mark the divisions on the three circles to show months, weeks, days, and hours. Using a protractor, divide the largest circle into 12 months of 30° each. You can now label the months from January through to December. Now divide each month into its number of days. Look at a calendar to see on what day January 1 falls this year and mark the 52 weeks of the year. Draw the markings in different colors for each month.

Phases of the Moon

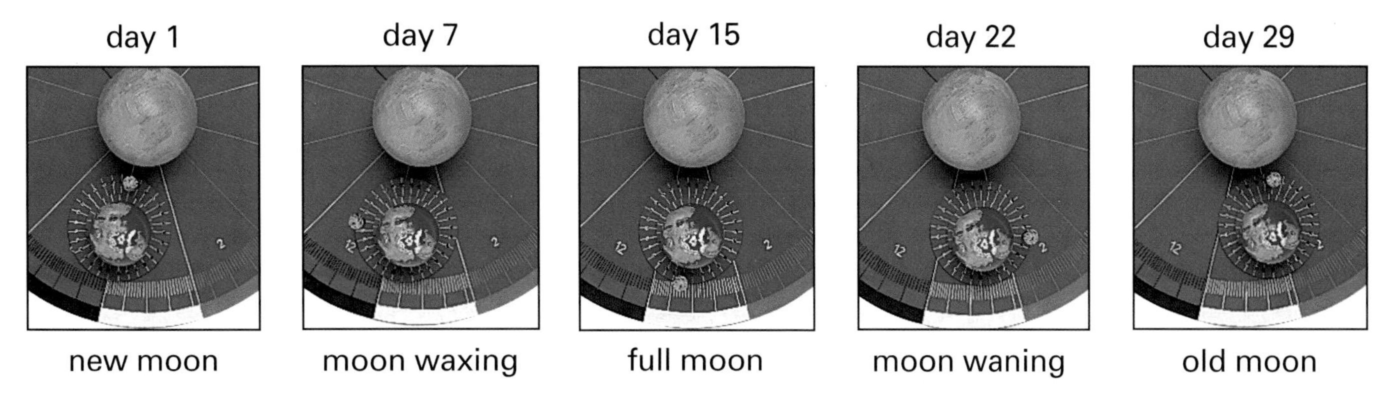

4 For the day lines, divide the middle-sized circle into 30 sections of 12°. As Earth goes around the sun, the moon also moves around Earth. It takes about 29½ days, a **lunar** month, for the moon to move around Earth back to the same position between Earth and sun.

Using your calendar

Check the date. Place Earth on the correct day and month on the calendar. Use a diary to find the "phase of the moon" on this day, and position the moon as above. As each day passes, move Earth and the moon on in their orbits.

5 The small circle shows how Earth spins. Divide this into 24 hours.

6 Glue the moon to the dowel. Punch holes through each of the day lines on the middle-sized circle so that you can move the moon around its orbit on the dowel.

7 Mount the disks and balls on the short length of cardboard tube as shown.

There is not an exact number of days in a lunar month, or lunar months in a year, so we divide the year into 12 uneven months. The names of the months come from the names the Romans gave them. March was once the first month of the year, which is why September, meaning "seventh," is now our ninth month.

10 Calendars

You know the date of your birthday, but do you know on which day of the week you were born? On which day of the week will your birthday be in 2021? Working out the day of the week on a particular date is a tricky mathematical problem. This 1,000-year calendar helps to solve the problem.

1 Using colored pens and a ruler, carefully copy the figures and lines onto the seven cardboard rectangles exactly as shown below and on page 11. Make sure you use red and black numbers on the large rectangles as shown. The red numbers on the third large rectangle show which years are leap years.

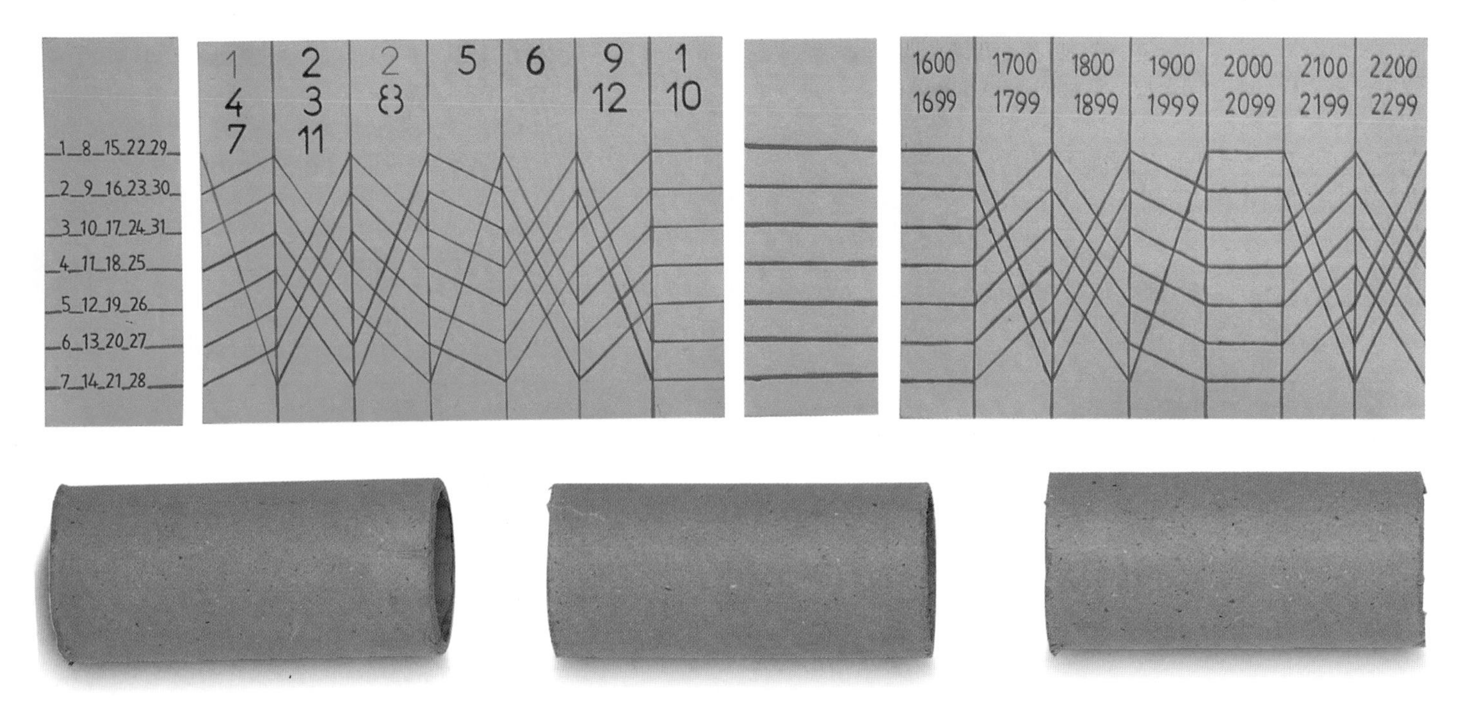

MAKE it WORK!

In a normal year there are 52 weeks and a day. The extra day means that if your birthday was on a Sunday this year, it will be on a Monday next year. A leap year moves your birthday forward two days. This calendar lets you find the day on any date in a 1,000-year period.

You will need

a ruler and pens
glue
a drill
thin balsa wood
thumbtacks
dowels
a 12" x 3" piece of plywood
scissors
three $2^1/_2$" diameter cardboard tubes
four 5" x 2" thick cardboard rectangles
three 7" x 5" thin cardboard rectangles

2 To make the dials, wrap the larger rectangles around the cardboard tubes and stick them in place.

3 Drill three holes in the plywood about 3" apart. Glue the dowels into them. Cut three balsa-wood circle caps and drill a hole in the center of each. Glue the caps to the tubes. Stand the tubes over the dowels, and pin the caps to the dowels so that the tubes turn freely.

4 Fix the four smaller pieces of cardboard to the plywood at either end and between the tubes, as shown. You may have to trim these so that the gaps between them match the column widths on the large cardboard pieces.

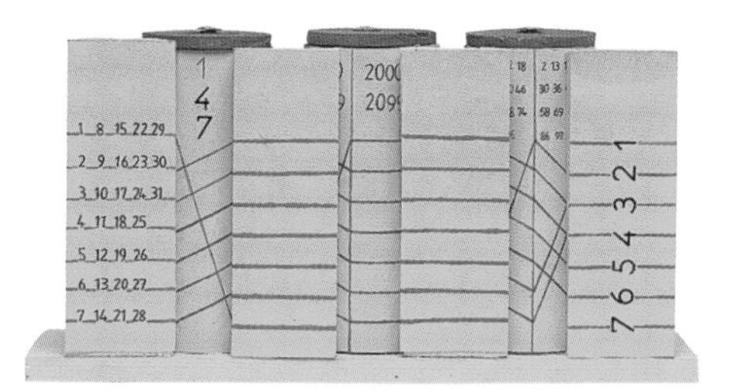

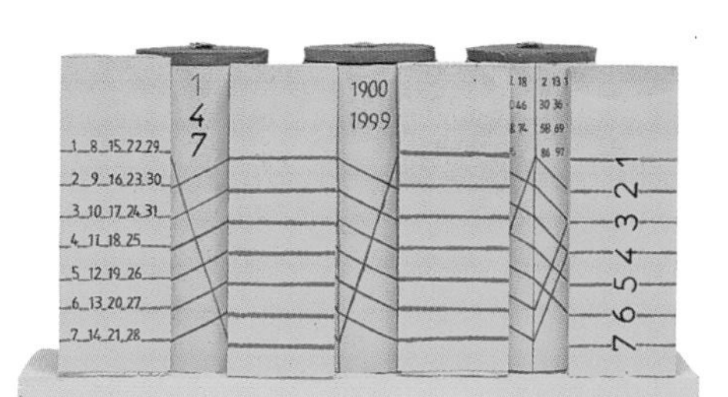

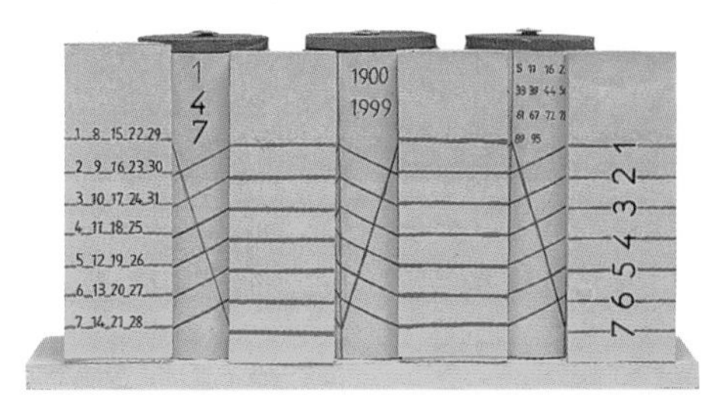

▲ Setting the dials

To find the day of the week of July 21, 1995, for example, turn the first dial to the number of the month (July=7). Turn the second dial to the **century** (1900-1999) and the third dial to the year (95). For leap years, use the red numbers on the first dial for January and February.

This calendar is based on the Christian calendar. The Muslim calendar is based on the lunar year of 13 lunar months. As the lunar year is 354.3 days long, the Muslim New Year moves steadily backward through the solar calendar.

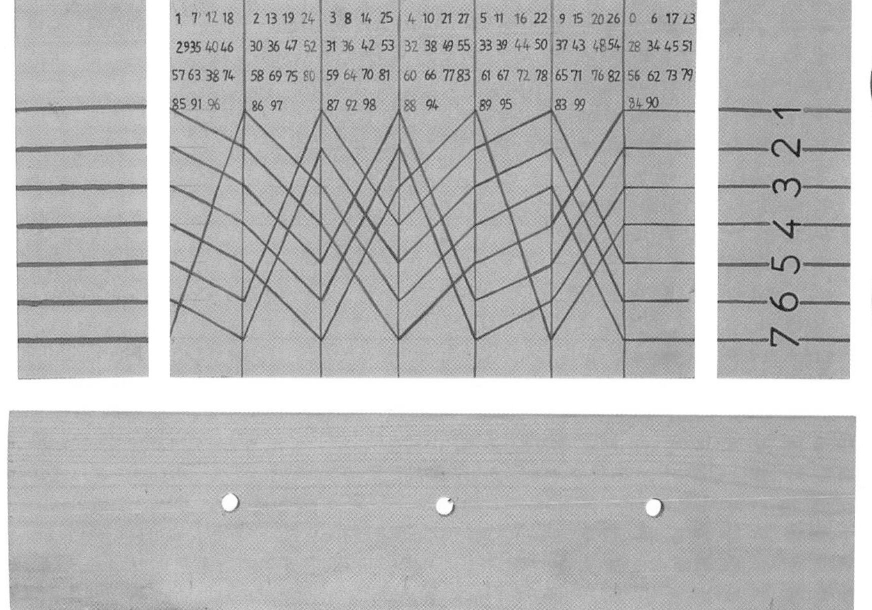

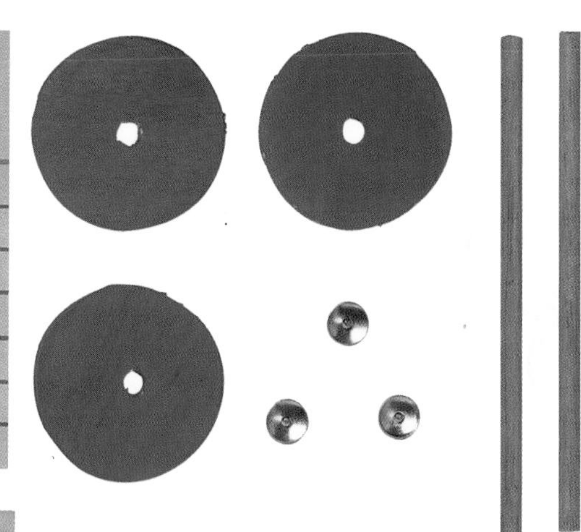

As the Earth takes 365 ¼ days to orbit the sun, we add a day to the calendar every fourth year to keep in step with the seasons. Leap years can be divided by four, for example, 1996.

Reading the dials

Now find the day of the month (21) on the left card. From this line, follow the horizontal and diagonal lines across the dials to the right card and you will discover that July 21, 1995, was the sixth day of the week—a Friday.

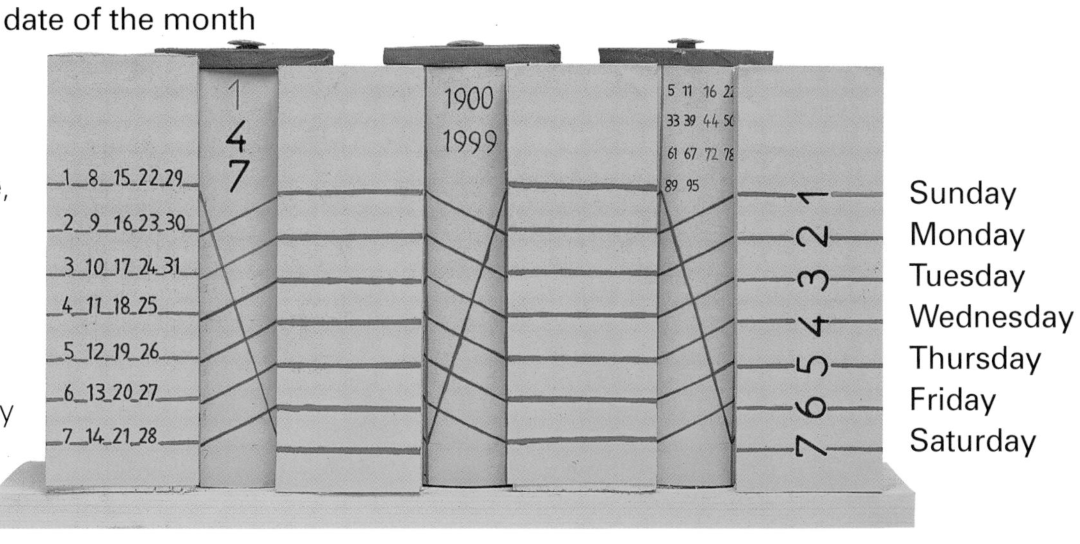

12 Longitude

When it's midday in Europe, it's midnight in Australia. The time of day depends on your position around the globe, or **longitude**. Longitude is measured by a series of imaginary circles that cross the equator at right angles and meet at the poles. Everyone on the same longitude, whether he or she is in the Northern Hemisphere or the Southern Hemisphere, sees the sun reach its highest point at the same time each day.

MAKE it WORK!

This model of Earth is a globe made by gluing together segments of Earth's surface along lines of longitude. It is not a perfect sphere because it is made from pieces of flat cardboard, but it gives a more accurate picture of Earth than a flat map. Longitude is measured in degrees east or west from the line of longitude that passes through Greenwich, England. This line is called the prime meridian and lies at 0°.

You will need

glue
paints
an atlas
scissors
thin cardboard

1 Draw a $16\frac{1}{2}$" line on a piece of thin cardboard. Now draw five lines at right angles to the long line, dividing this line into six equal sections. These lines, **a**, **b**, and **c** below, should be $2\frac{3}{4}$", 2", and $1\frac{1}{2}$" long. Join the end points together, and add six tabs to make a segment like those shown below. Cut out the segment. Cut 11 more segments the same size.

2 Draw the outlines of the continents on your segments. Use the model shown as a guide.

3 Paint the land and oceans in appropriate colors. Mark the lines of longitude down the center and along one edge of each segment. Paint the equator on in red as shown at right.

4 Fold and glue each of the segments to build up the globe, as shown above.

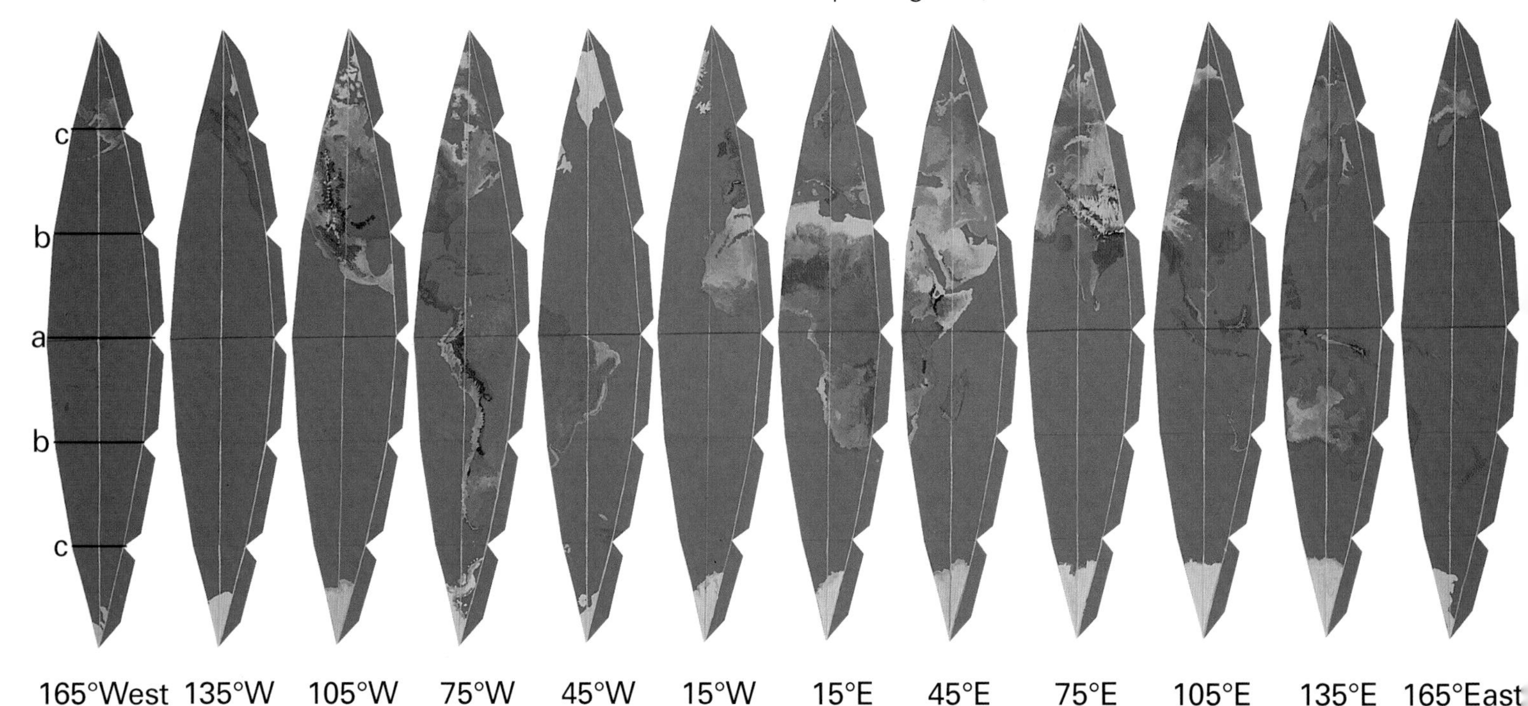

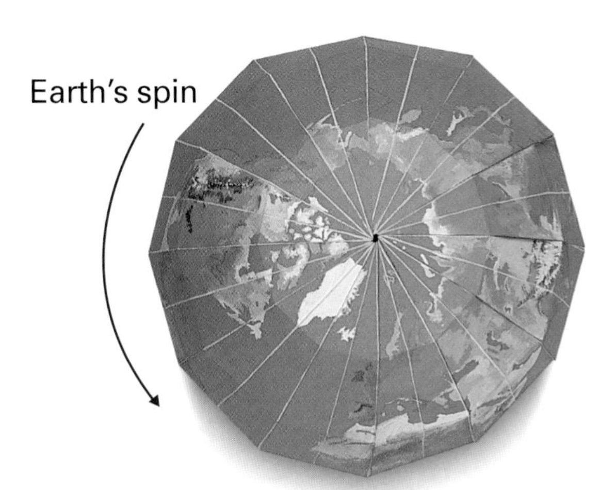

a view from above the North Pole

The 24 lines on your model are spaced 15° apart. Earth turns a complete circle of 360° in 24 hours, so in one hour it turns through 15°. Looking down on the North Pole, one would see Earth spinning **counterclockwise**, so midday on the 15°E line of longitude occurs one hour earlier than on the prime meridian.

The longitude of New York is 74° west. The sun reaches its highest point above New York just less than five hours later than it does above London.

If you set your watch to Greenwich Mean Time (GMT), it shows 12 noon when the prime meridian is facing directly toward the sun.

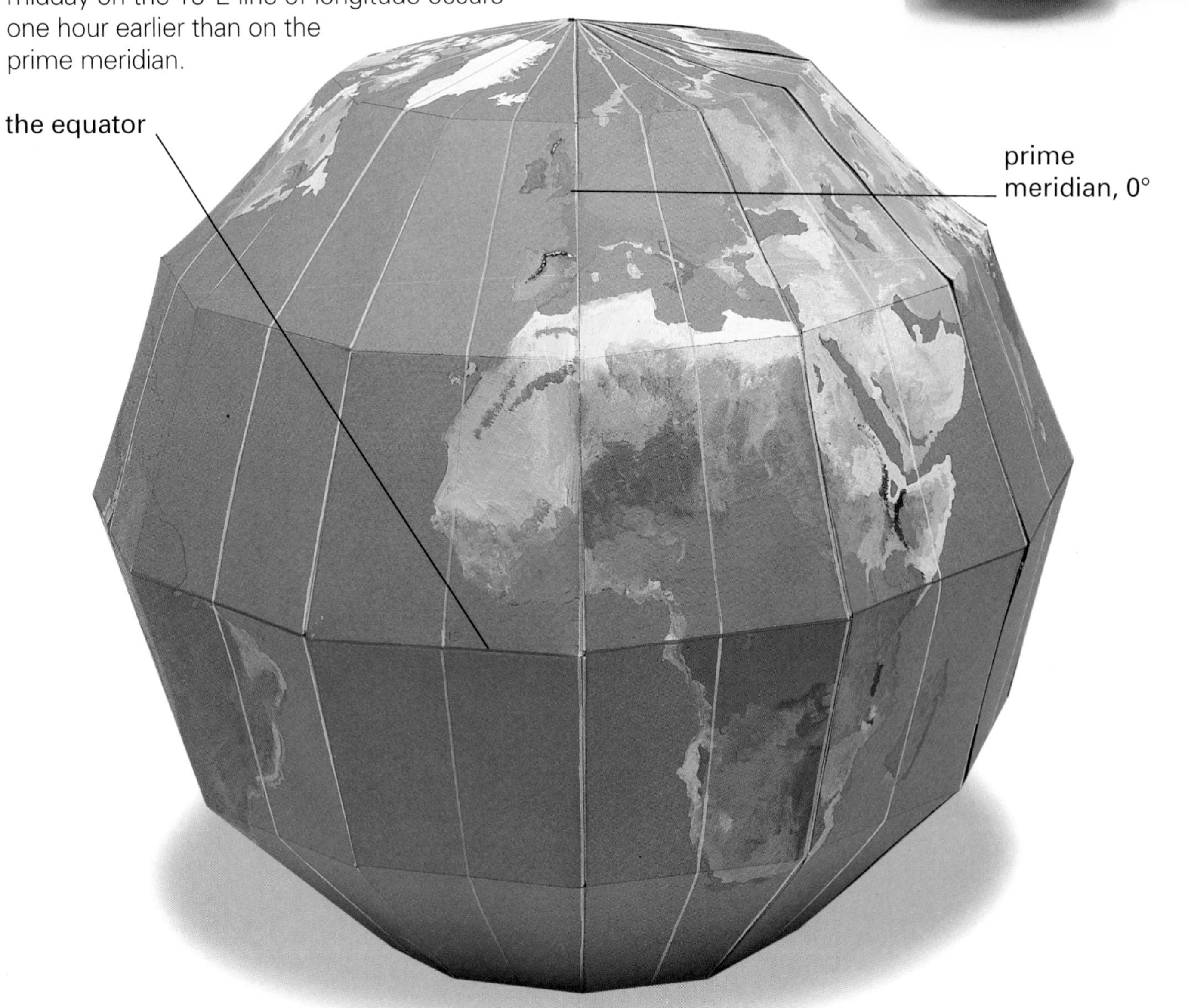

14 Time Zones

Have you ever flown to another country on vacation? Just before you land the pilot usually tells you what the local time is so you can set your watch. The world is divided into 24 different time zones. As you move east or west into a new time zone you have to set your watch one hour forward or back.

MAKE it WORK!
In theory, time zones should follow lines of longitude, with one hour change every 15°. But in practice this would mean that people on opposite sides of a city through which a zone line passed would have their watches set an hour apart. Countries have agreed that time zones should follow the borders between countries. Some large countries, such as the United States, Russia and Australia, have several time zones within their borders.

This time zone cylinder will help you check the number of hours time difference between places at different longitudes.

You will need

glue
an atlas
paints
thin cardboard
a wide cardboard tube

1 Cut a rectangle of thin cardboard so that it just wraps around the cardboard tube. Make the width of the rectangle $1^1/_2$" less than the height of the tube.

2 Divide the thin cardboard into 24 equal sections and copy the time zone map below. The different colors represent the time zones. The rectangles along the top correspond to the colors in the time zones.

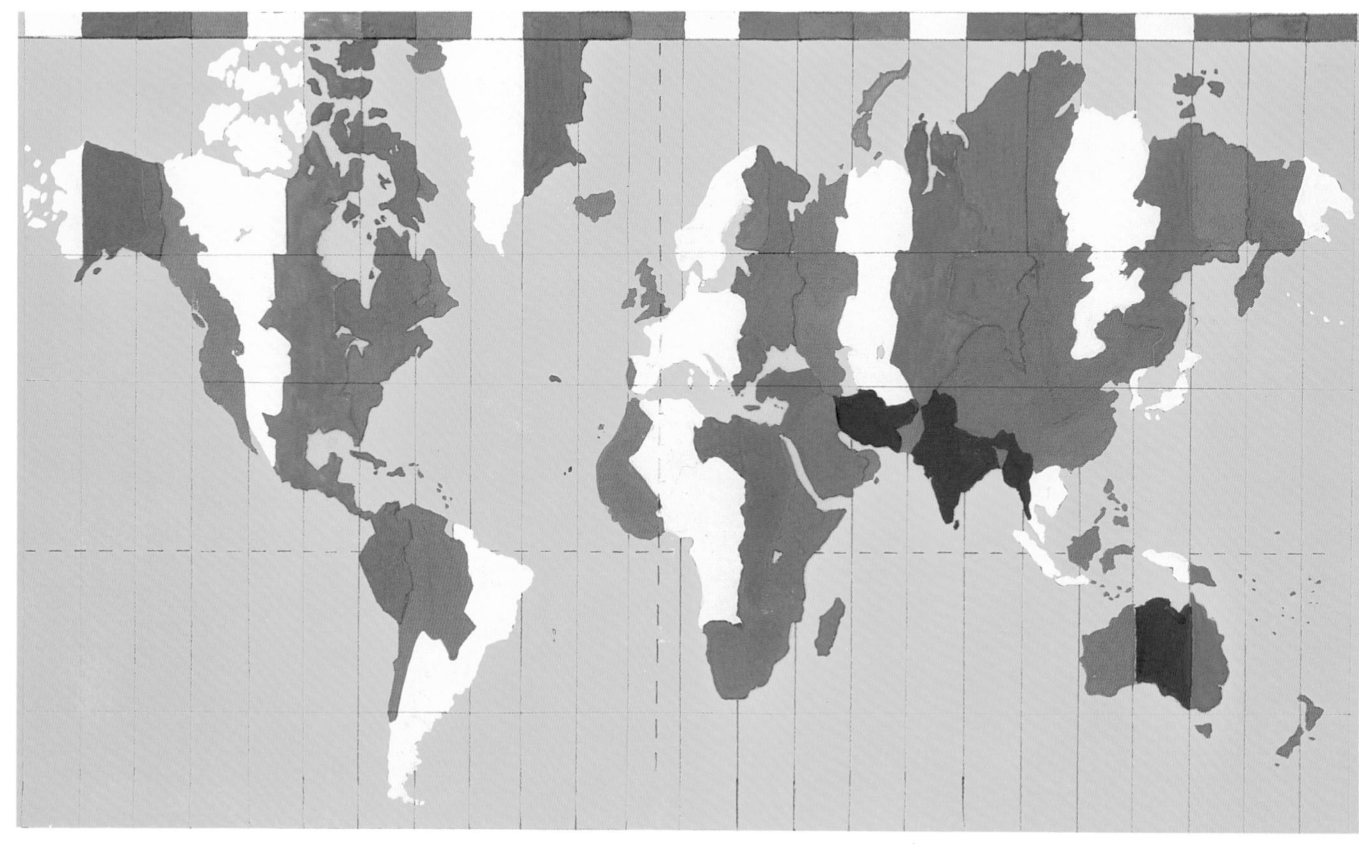

3 Wrap the map around the tube, leaving a $^{3}/_{4}$" gap at each end. Glue it in place.

4 Cut two thin cardboard disks $^{1}/_{16}$" larger than the diameter of the tube. Now cut two $1^{1}/_{2}$" wide thin cardboard strips that just wrap around the tube. Cut triangular glue tabs along one edge of each strip.

5 Glue the strips to the edges of the disks to make caps to fit over the ends of your cylinder.

6 Cut a $^{3}/_{4}$" strip of thin cardboard that just wraps around the tube. Use a ruler to divide it into 24 equal time zones. The divisions should measure the same as the colored rectangles along the top of the map. Label the sections as below. Glue it onto the top cap.

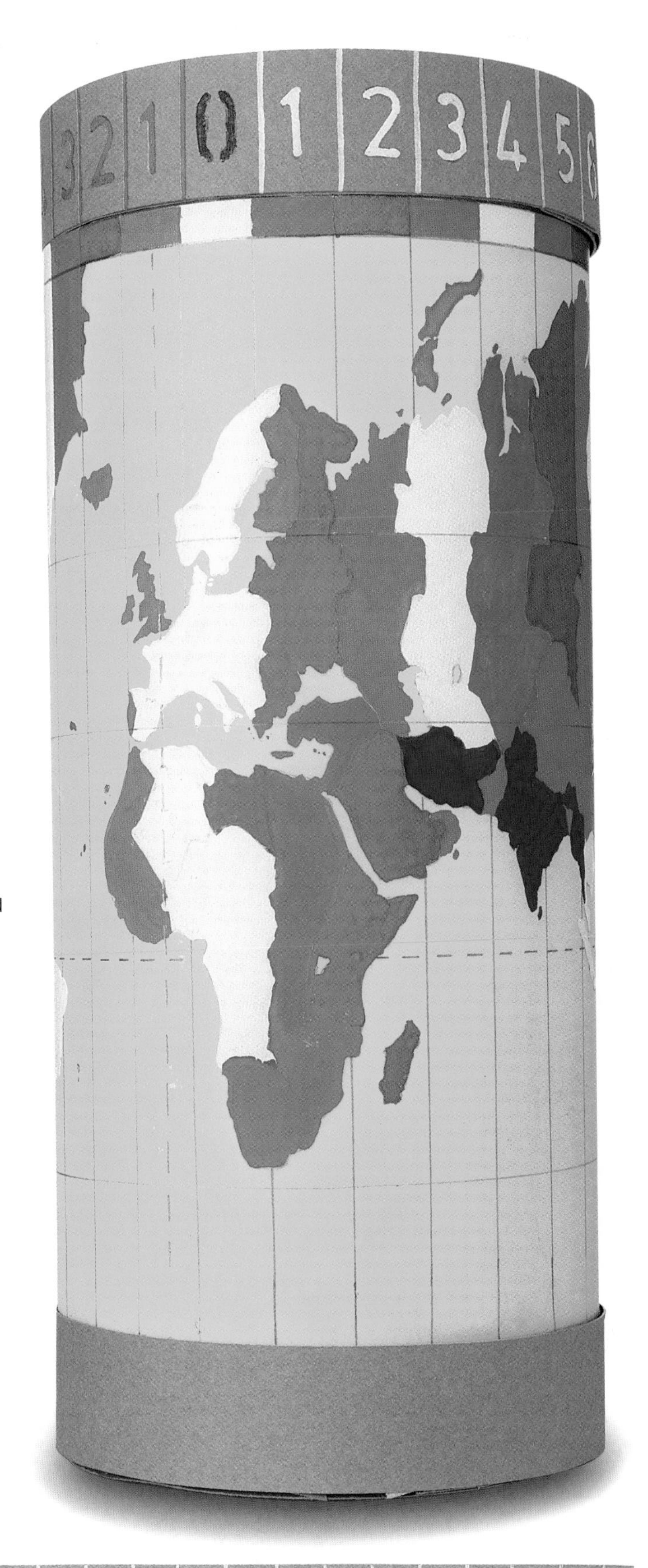

Using your time cylinder

First check the color of the time zone where you live. Turn the cap so that zero is lined up with a rectangle of this color above your time zone. You can then read off the time differences in other time zones. The white numbers show the number of hours ahead of your time zone and the red ones the number of hours behind.

▶ This time cylinder is set for someone living in Paris, France, in the white zone. In England the time is one hour behind. In the green zone of Saudi Arabia the time is two hours ahead.

If you travel west from one zone into another then you must put your watch back by an hour. When you travel east across time zones your watch goes forward. The countries shown in purple chose to set their clocks half an hour ahead or behind the neighboring time zone.

11 10 9 8 7 6 5 4 3 2 1 0 1 2 3 4 5 6 7 8 9 10 11 12

Latitude

You must reset your watch if you make a long journey east or west and change your longitude. Traveling north or south to a different **latitude** does not affect the time. But it does change the number of daylight hours and the height of the sun in the sky at noon.

MAKE it WORK!
With this quadrant you can find your latitude by looking at stars at night. You need to know your latitude to make the sundials on pages 18-21.

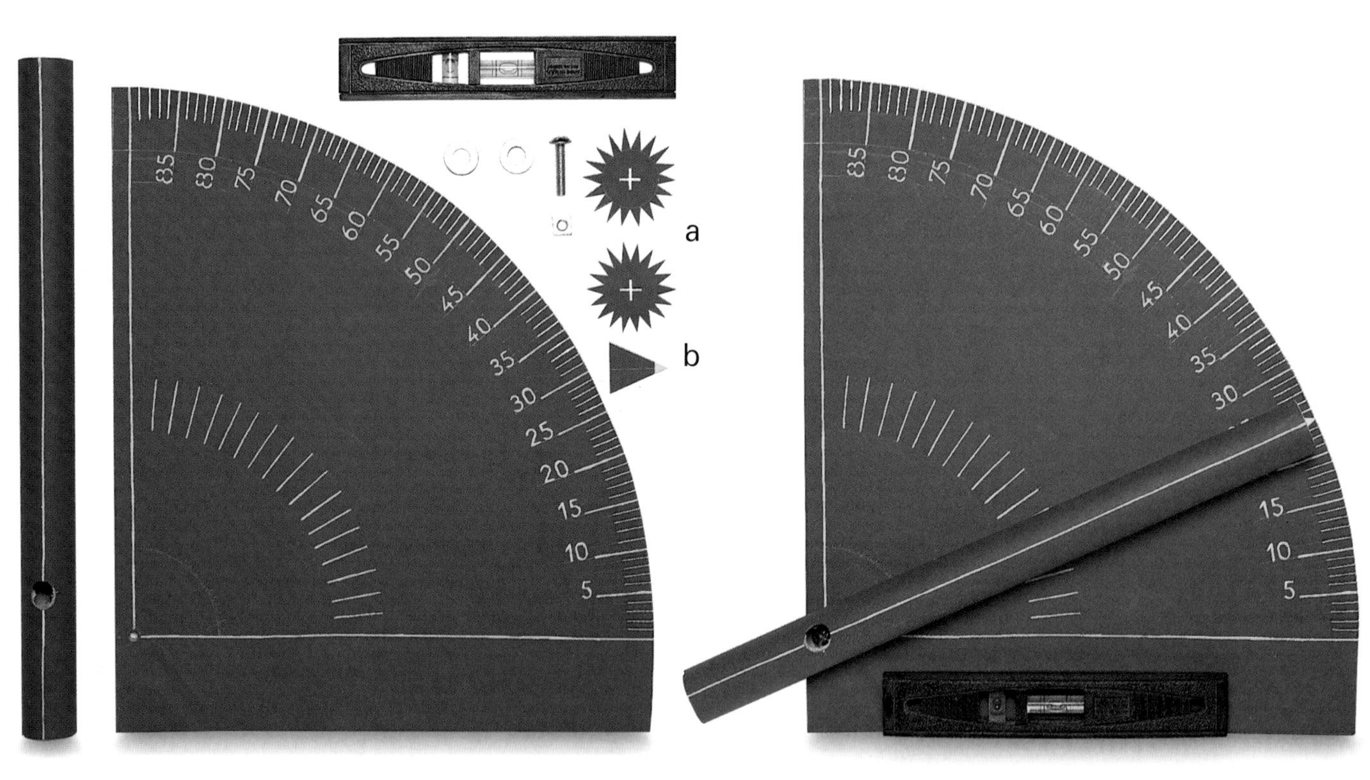

You will need

a level	pens
a protractor	cardboard
plywood 12″ square	glue and scissors
a nut, a bolt, and washers	20″ cardboard tube

1 Mark a quarter circle (a quadrant) with **radius** 16″ on a sheet of cardboard. Cut out the shape, leaving a 3/4″ border along one edge and a 2″ border along the other, as shown.

2 Use a protractor to make scale marks at intervals of 1° around the edge of the quadrant. Label the marks at 5° intervals.

3 Cut out two cardboard caps with tabs, **a**, to cover the ends of the tube. Cut sighting slots in the center of the circles. Glue caps in place.

4 Drill a hole for the bolt 4″ from one end of the tube. Enlarge the hole on one side of the tube so that the bolt head can pass through. This will stop the bolt blocking light inside the tube.

5 Bolt the sighting tube to the quadrant so that it will turn around the scale.

6 Cut a pointer, **b**, and glue it onto the tube so that it points to the lines on the scale, as above. Glue the level in place so that it lines up with the bottom edge of the quadrant.

7 Outside, place plywood base on a stool or brick tower, and stand your quadrant on it.

You can find your latitude by pointing the sighting tube towards the **celestial** pole.

a planisphere

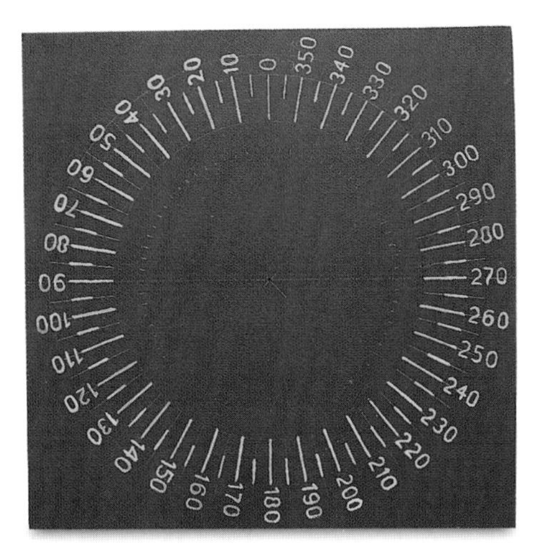

Watching the "moving" stars
If you fix on another star in the sky, you will find that it seems to move around the celestial pole through the night. Divide the plywood into degrees, as shown at left. Position the quadrant across the middle of the plywood. You can now pinpoint your star in the sky with a horizontal angle and a vertical angle. Track your star across the sky at hourly intervals.

Positioning your quadrant
The celestial poles are positioned directly above Earth's poles. These are the only points in the night sky that do not seem to move as Earth spins. Look at a star chart or a planisphere to help you locate the celestial pole. This will be the North Star if you live in the Northern Hemisphere and the South Celestial Pole if you live in the Southern Hemisphere.

Turn the quadrant so that the curved edge faces in the direction of the pole. Check your level to be sure that the quadrant is horizontal. Look through the tube and turn it to line up the celestial pole in the sighting slots. Read the angle of the tube from the scale. This is equal to your latitude.

18 Sundials

The first clock was probably just a stick stuck in the ground. Sundials have been used for thousands of years to tell the time. As Earth spins, the sun seems to move across the sky. The stick's shadow moves with the sun, and you can tell the time of day from the direction in which the shadow points.

You will need

- a protractor
- two strips of wood
- stiff cardboard for gnomon
- thin cardboard and scissors
- a square piece of plywood
- pens
- a ruler
- a watch
- a stencil
- a compass

1 Cut your cardboard gnomon. For the gnomon to point in the direction of Earth's axis, use a protractor to make the angle equal to the latitude of the place you will use the sundial (for example, 41° for New York).

MAKE it WORK!
The shadow of a vertical pole changes length as the sun rises and falls in the sky. At sunrise and sunset the shadow is very long. It is shortest at noon when the sun reaches its highest point. This sundial has an angled pointer called a gnomon. Early sundial makers discovered that if they angled the gnomon to point in the same direction as the Earth's axis, then the length of the shadow changed less and the sundial would work throughout the seasons. The spaces between the hour marks on the dial are also more even.

2 Cover the plywood with thin cardboard. Using the wood strips as supports, glue the gnomon in place, as shown.

◀ a gnomon for use in London

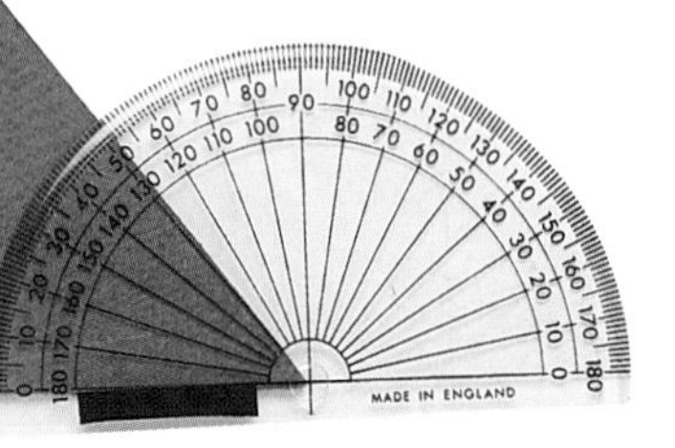

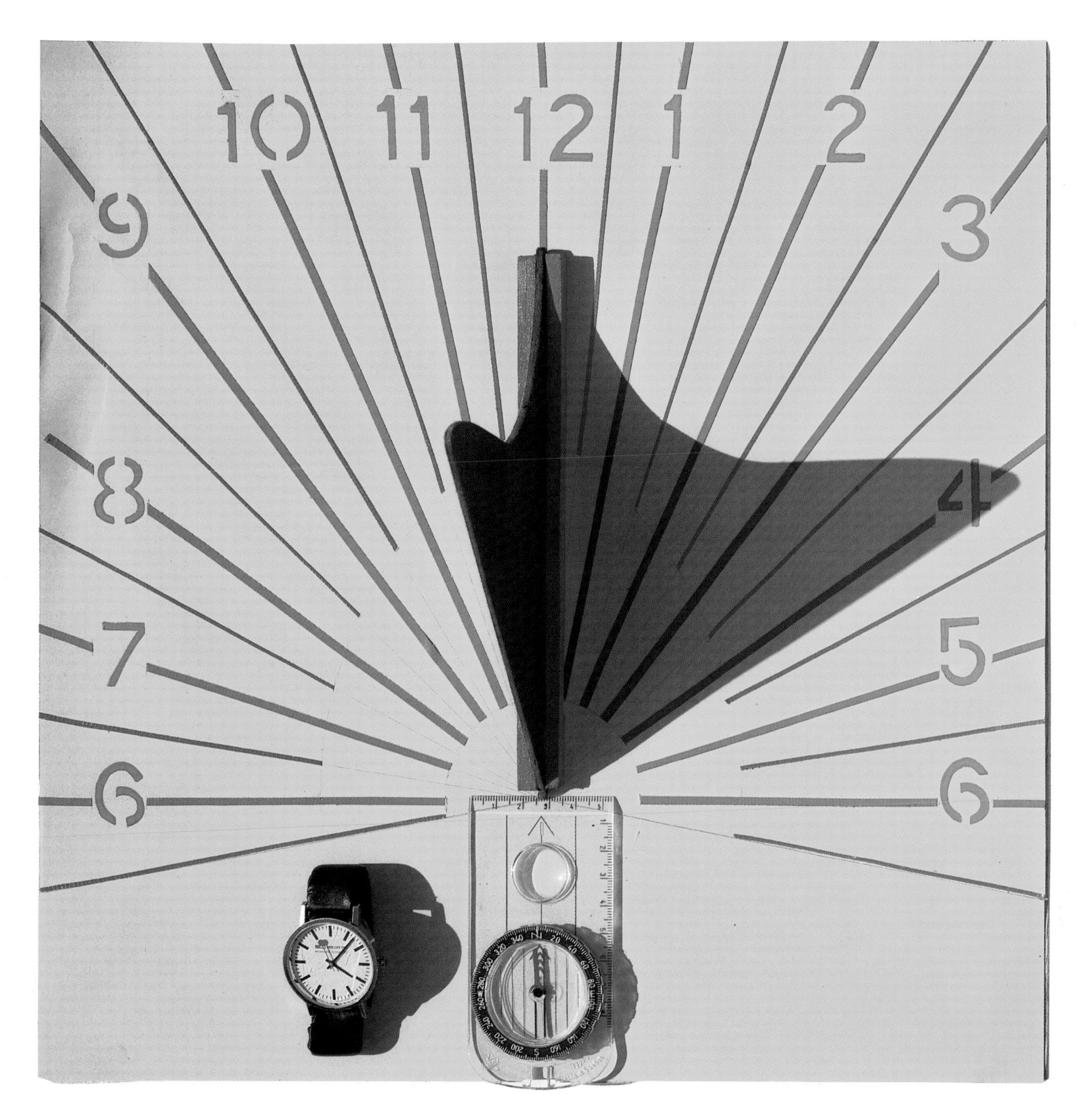

3 Stand your sundial on a flat surface in the sun. You can now **calibrate** your sundial. Use a compass to line up the dial. The straight edge of the gnomon must face south if you live in the Northern Hemisphere, or north if you live in the Southern Hemisphere. Check your watch and mark the position of the shadow every half-hour.

4 Check that your sundial is calibrated correctly by using it to tell the time on the next sunny day.

We normally think of noon as the time when the sun reaches its highest point. This depends on your longitude. If you have calibrated your sundial with a watch, you may find that the sun is not at this point at midday. This is because the time on your watch depends on your time zone. All people in one time zone set their watches to the same time, whatever their longitude.

20 Universal Sundials

Most sundials are fixed in one place, with the pointer and the spacing between the hours set for a particular latitude. If the dial is moved to a different latitude, then it does not tell the time accurately. But this universal sundial can be adjusted to tell the time anywhere on Earth!

You will need

a ruler	a compass
a watch	thin cardboard
balsa wood	thick cardboard
glue and tape	a protractor and scissors

1 Cut the balsa-wood shapes. First cut the two yellow pieces for the frame. Cut slots as shown so they fit together to make a cross. Cut three semicircles (one yellow and two red) with a flat edge of 4″. Cut one 4″ x $^1/_8$″ strip.

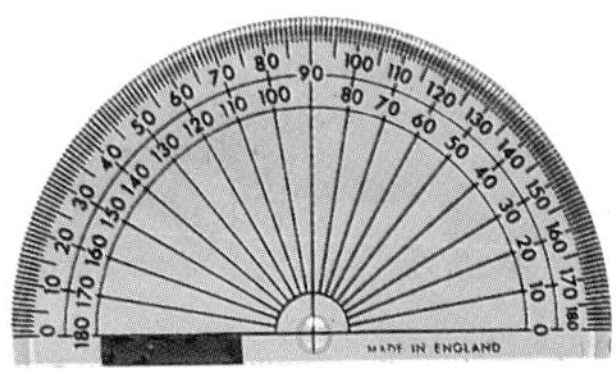

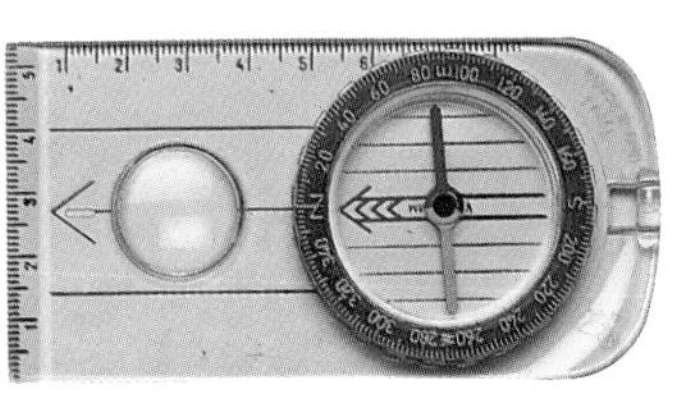

2 Cut the thin, red cardboard pointer, as shown below left. The diameter of the inner circle should be 5″.

11 12 1
10 2
9 3
8 4
7 6 5

month dial

2″

7 8 9 10 11 12 13 14 15 16 17 1

8″ time dial

11 12 1
10 2
9 3
8 4
7 6 5

month dial

3 Cut the time dial and month dials. With a pen, divide the time dial into 12 equal columns. Divide each column into four equal strips with a pencil.

pointer

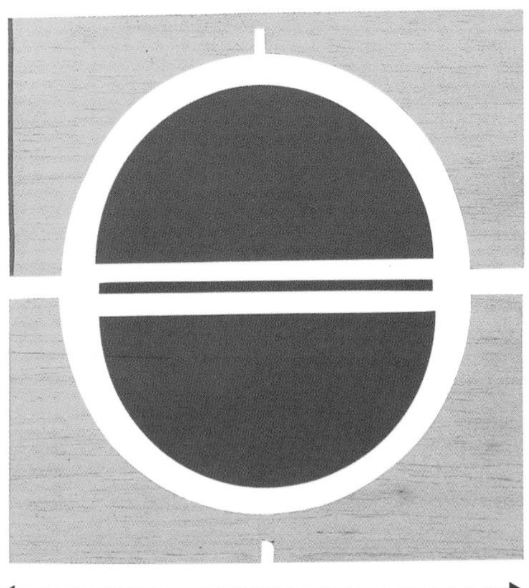

6″

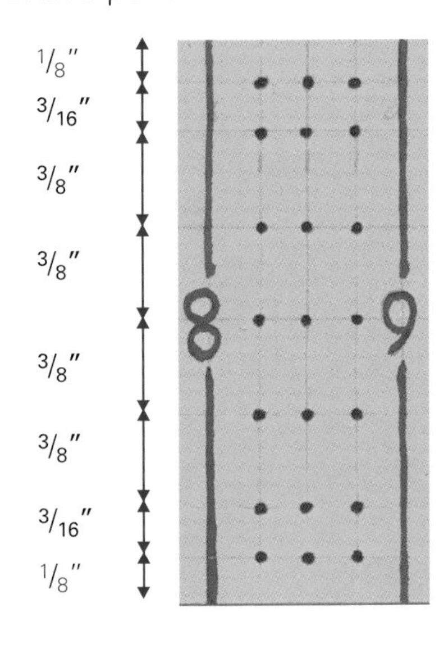

MAKE it WORK!

The pointer on this sundial is a long slot that lets the sun's light through. A line of light falls on a curved scale to show the time. If you move to a different latitude, you can adjust the angle of the pointer. This means that the line of light can be made to move across the scale at the same rate (15° per hour) wherever the dial is on Earth.

4 Label the dials as shown. Use the dimensions shown to position your dots for each column. The numbers on the month dials must line up horizontally with the dots on the time dial.

5 Glue the three semicircles of wood, the narrow strip, and the protractor together to make the base of the sundial.

Reading the dial

The shadow cast by the central slit moves across the the time dial. The vertical line shows you the time. On the sundial below it is about 10 minutes past four. The horizontal slit in the pointer is for the month. Follow the dots across from the shadow of the slit and read off the numbers from the month dial. The dial will give you two choices for the month. On the sundial below, it is either July (7) or May (5).

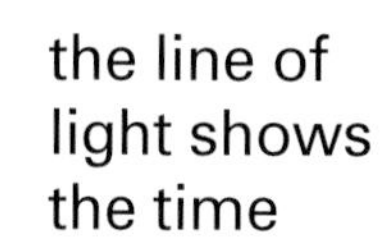

6 Slot the two arch-shaped pieces of wood together at right angles to make the bowl-shaped frame.

7 Bend and glue the dials onto the frame, as shown above.

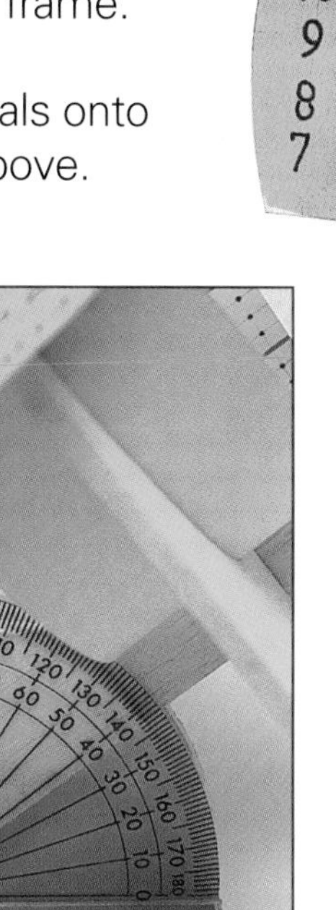

Setting the dial

Slot the frame between the upright semicircle and the protractor so that the corner point is at the center of the protractor. Slide the frame to the angle of your latitude and tape the frame in place. Now take your sundial into the sun and use a compass to position it correctly. The front must face south in the Northern Hemisphere or north in the Southern Hemisphere.

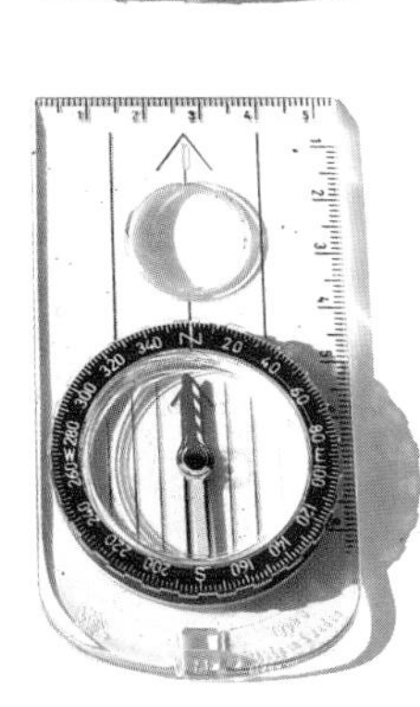

22 Time at Night

At sunset the shadow disappears from a sundial. It won't tell you the time again until sunrise on the following day. But Earth continues to spin. At night you can tell the time from the way the stars appear to move across the sky.

You will need

- a star chart
- a protractor
- a large washer
- a drawing compass
- a metal stud with a hole in the middle
- paints
- string
- scissors
- cardboard

Making the dial

1 Draw a large, a medium, and a small disk on a piece of cardboard as shown. Add a handle to the large disk, three pointers exactly in the positions shown to the medium disk, and a "star arm" to the small disk. Cut out the shapes.

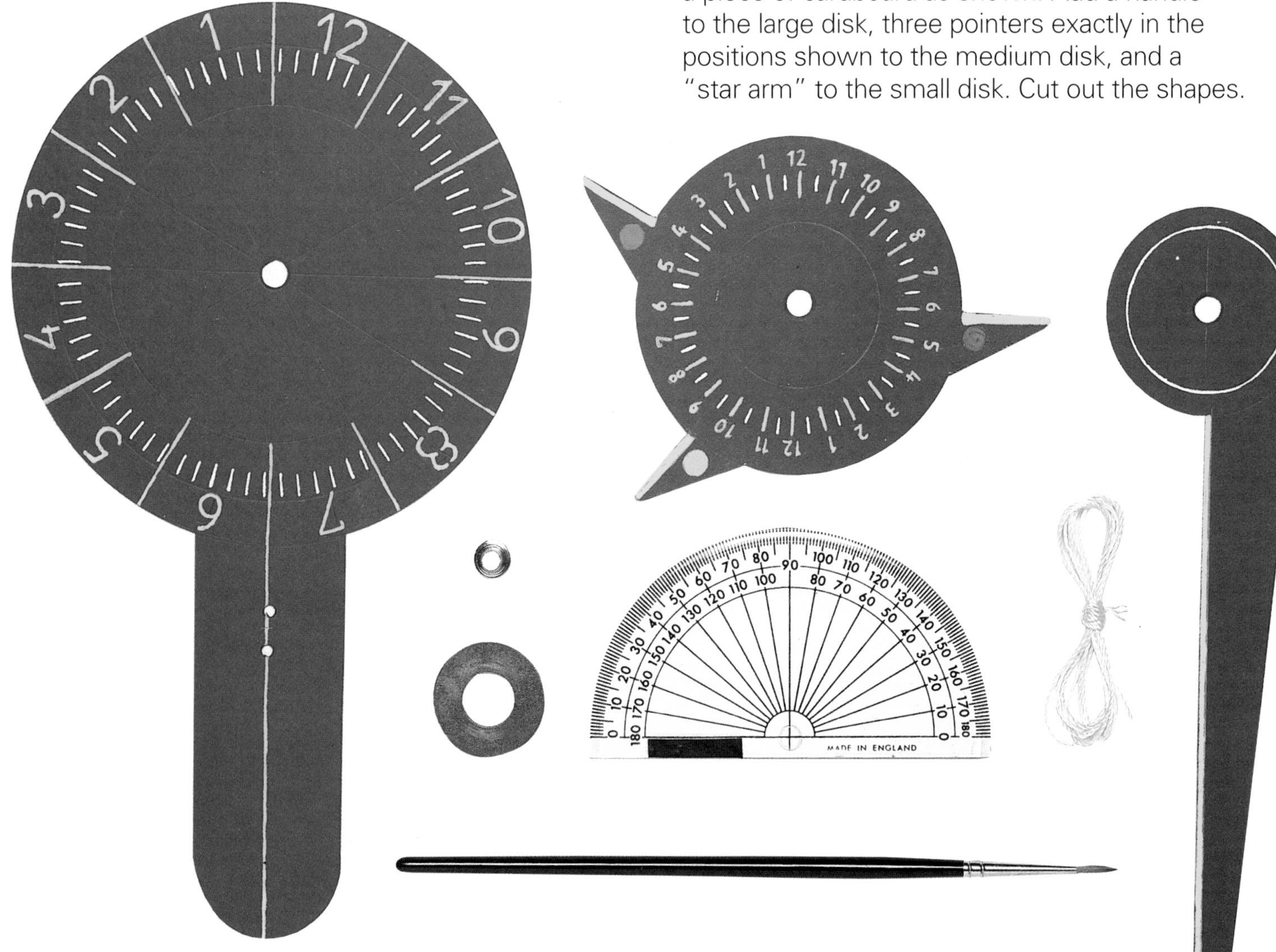

MAKE it WORK!

The spinning of Earth seems to make the sun move during the day and the stars move at night. The stars turn around the two celestial poles. You can make a device that you can line up on the stars to tell the time. You will be able to use it when your sundial has gone to sleep!

2 Use a protractor to help you label the large and medium disks. The yellow divisions on the large disk are the 12 months. The white divisions are intervals of five days. The medium-sized disk is divided into 24 hours. Make sure your hour labels line up with the three pointers, as shown above.

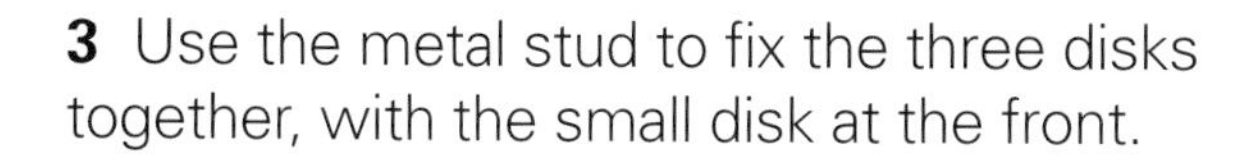

3 Use the metal stud to fix the three disks together, with the small disk at the front.

4 A plumb line will help you keep your device vertical. Hang a washer from the handle to make your plumb line.

Setting the dial

The three parts of your device are the date dial, the time dial, and the star arm. Line up one of the three pointers on the time dial with today's date. The three pointers are for three different stars:

red—Deneb
green—Dubhe
yellow—Capella

star arm
pointer for Deneb
date dial
time dial
pointer for Capella
pointer for Dubhe

◀ With the star arm pointing at Dubhe, the time shown here on May 24 is 11 o'clock.

1 Choose one of the three stars and move its pointer to the date. Use a star chart to help you find your chosen star in the night sky.

2 Now hold up your device and look through the hole in the center toward the North Star. Use the plumb line to check that the handle is vertical.

3 Without moving the time dial, rotate the star arm until it points toward your chosen star.

4 The place where the marked edge of the star pointer crosses the time dial shows you the time.

You can use this device when the stars are partly hidden by cloud as long as at least one of the three pointer stars is visible.

24 Candle Clocks

Imagine you are a medieval monk or nun. You must spend certain hours each day praying. Clocks and watches have not been invented, and you cannot interrupt your prayers to go outside to look at the sun or the stars. How can you tell the time? The solution is to light a candle. It gives you the light to see by and tells you the time. You need to choose a candle that will burn for a set number of hours.

1 Spread the gravel on a fireproof surface such as a metal tray. Push each of your candles into a ball of clay on the gravel.

2 Light one candle of each type only, and use the stopwatch to check how far it burns in set times—every **minute** for the small candles and every 15 minutes or 30 minutes for the larger candles. Use the tape to mark the unlit candles at the timed intervals so that you can use them later as candle clocks.

Warning—do not light too many candles at once. The heat may cause them all to melt. Never leave a lit candle unattended.

MAKE it WORK!

A candle that is the same width from top to bottom burns at a steady rate. Fat candles usually burn more slowly than thin ones. You can test different-sized candles to see how useful they are as candle clocks. Remember to ask an adult to help you with this experiment.

You will need

- gravel
- matches
- thin strips of tape
- a collection of different candles
- clay
- a stopwatch

Chiming candle clocks

Mechanical clocks often chime a bell to mark the hours. This clever candle clock makes a sound at regular intervals, too.

You will need

- yarn
- gravel
- a watch
- matches
- a foil dish
- 12 metal washers
- a wax taper (a long thin candle)
- tape
- a brick

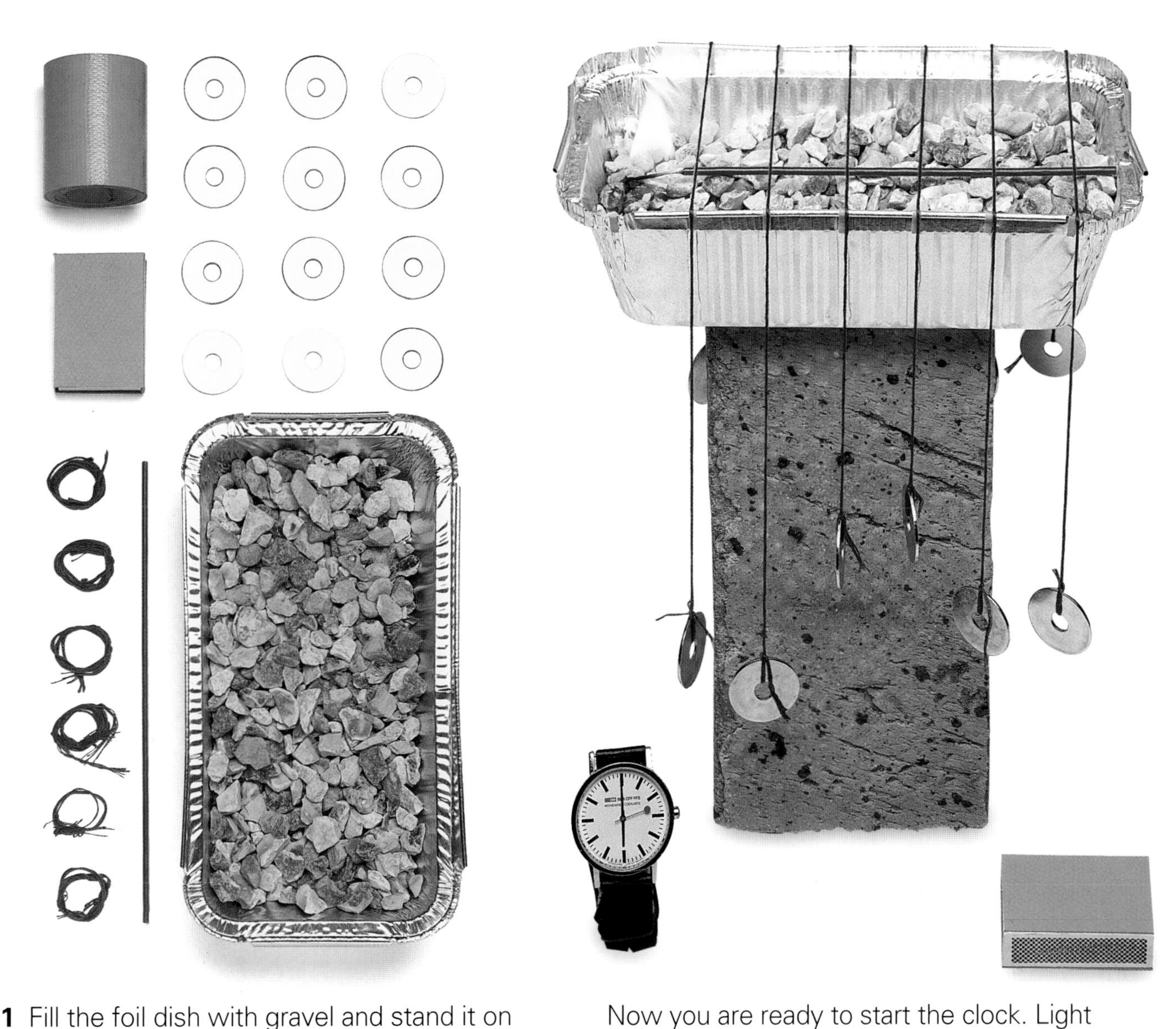

1 Fill the foil dish with gravel and stand it on the brick. Mark six evenly spaced lines along the edge of the dish with thin strips of tape. Now lay the taper along the gravel.

2 Cut six lengths of yarn. Tie a washer on each end of each length of yarn.

3 Hang the yarn lengths over the markers on the dish. The washers mustn't touch the ground.

Now you are ready to start the clock. Light the taper with a match. As the taper burns through each piece of yarn, the washers will drop with a clang. Use a watch to check the time between each chime of your candle clock.

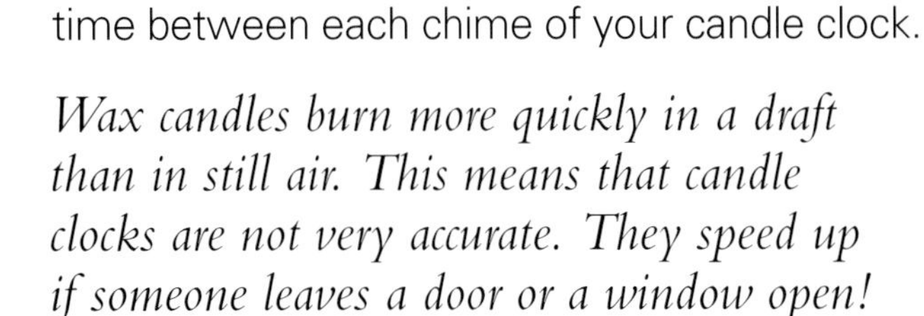

Wax candles burn more quickly in a draft than in still air. This means that candle clocks are not very accurate. They speed up if someone leaves a door or a window open!

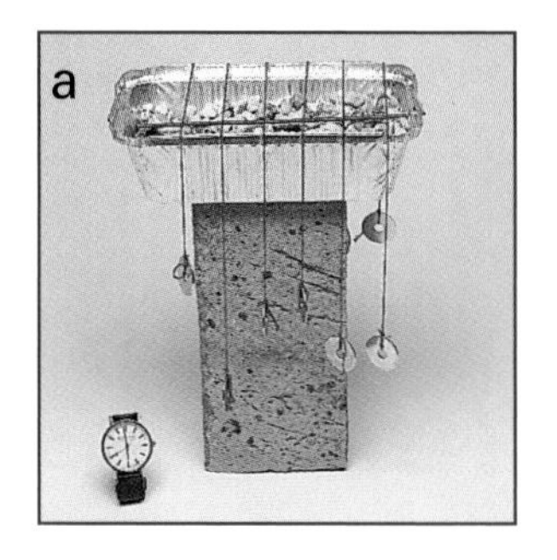

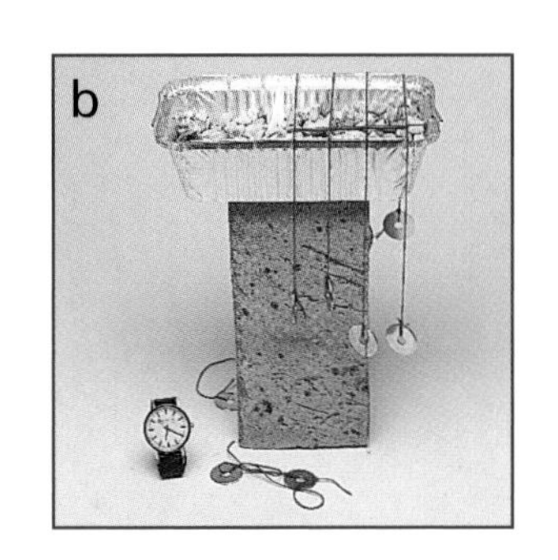

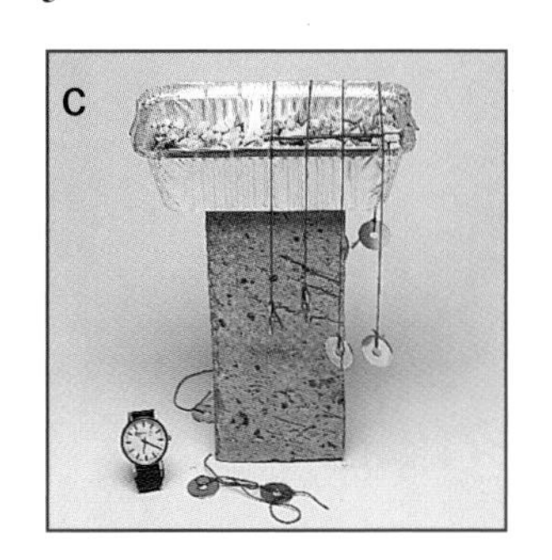

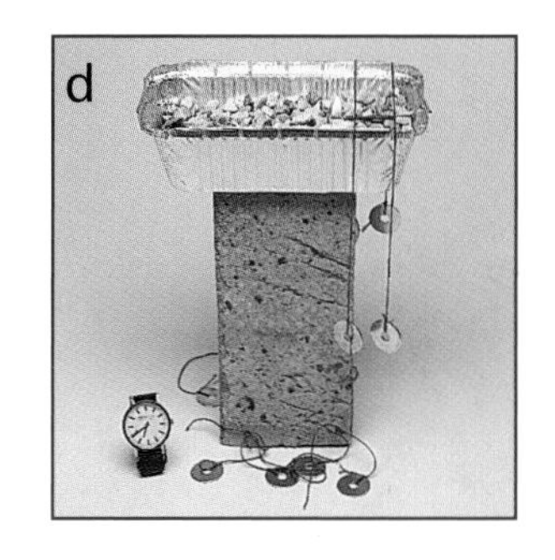

26 Water Clocks

The drip, drip, drip of water from a leaking tap can drive you crazy! Because the timing of the drips is so even, you know just when each one is about to come. But the regular drip of water through a small hole can be useful, too. You can put it to work to make a water clock.

You will need

- a tall glass
- a large cork
- a plastic tube
- a short dowel
- a plastic funnel
- a plastic drinking straw
- water and food coloring
- a thin strip of wood 4″ x 1″
- a piece of wood $4^3/_8$″ x $1^1/_2$″ x $^3/_8$″
- two pieces of wood 10″ x 2″ x $^3/_8$″
- three pieces of wood 12″ x 2″ x $^3/_8$″
- glue
- clay
- a pin
- a drill
- thick cardboard

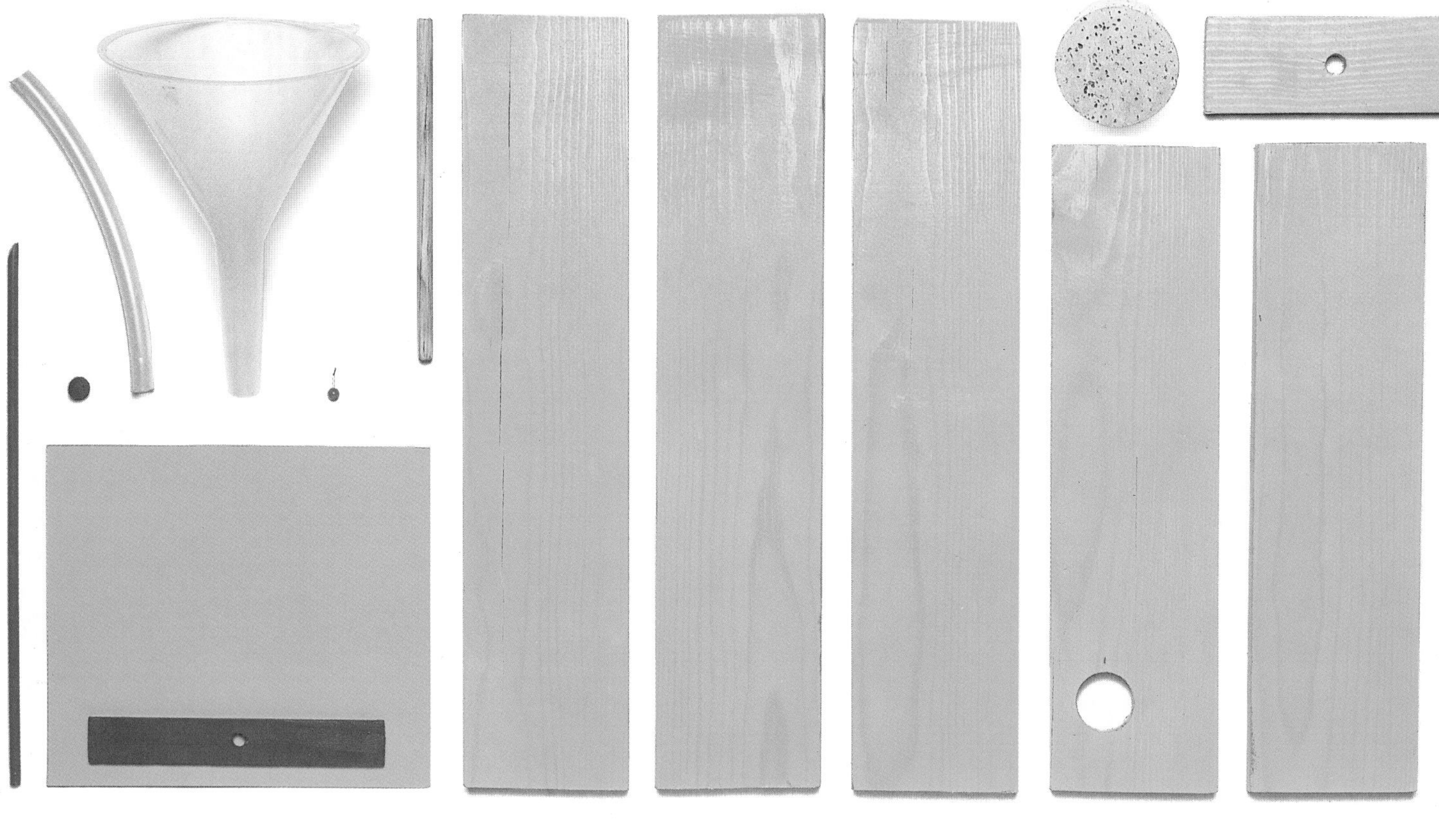

MAKE it WORK!

In this water clock, water from the funnel drips through a small gap at the end of the plastic tube. The water drops are collected in a tall glass. As the water fills the glass, it lifts a cork float. This moves a pointer across a dial. Because the water level increases steadily as time passes, the position of the pointer on the dial shows how long the water has been dripping. By calibrating the water clock with an accurate watch, you can use it to measure the passing hours.

1 The water clock frame is made from the various pieces of wood. Ask an adult to help you cut these to size. The dimensions given are a guide only. You may need to adjust these sizes to fit your particular funnel and glass.

2 Drill a large hole near one end of the top piece of the frame. This is where you rest the funnel.

3 Drill a hole just larger than the diameter of the dowel at the center of the short crosspiece that fits inside the frame.

4 Glue the frame together, as shown below. The crosspiece must be positioned below the funnel hole. Let dry.

5 For the scale, glue a piece of cardboard to the side of the frame. Do not make the scale marks at this stage.

6 Drill a hole the same diameter as your dowel in the center of the thin red wooden strip. Push one end of the dowel into the hole and glue it in place.

7 Make a hole in the top of the cork. Pass the free end of the dowel through the hole in the crosspiece and push it into the cork. Slide the glass into place with the cork inside it.

8 Now fix the plastic tube to the funnel and partly plug the open end with clay. Fill the funnel over a sink and adjust the clay plug until you get a steady slow drip. Then place the empty funnel and tube in the frame.

9 Pin one end of the straw to the frame so that it rests over the wood strip. Make sure the straw pointer will move freely as the strip rises.

Over 2,000 years ago the Greek inventor Ctesibius built water clocks that rang bells and operated puppets. But these were not the first water clocks. Water clocks, or "clepsydras," had been invented in Egypt at least 1,000 years earlier.

Calibrating your clock

First fill the funnel with colored water. Use a watch to check the time, and at regular intervals mark the position of the pointer on the scale. Depending on how you have adjusted the rate at which your clock drips, it may be suitable for measuring minutes or hours. Once your clock has been calibrated, refill the funnel and experiment to see whether the clock keeps accurate time.

28 Pendulum Clocks

A good clock must tick at a steady rate. In 1581, when he was 16, the great Italian scientist Galileo Galilei sat in the cathedral at Pisa watching a hanging lantern swing to and fro. He realized that the lantern was a pendulum and that each of its swings took exactly the same amount of time.

You will need

- a drill
- dowel
- stiff cardboard
- glue and string
- two thick cardboard tubes
- a thumbtack
- a metal washer
- a fishing line reel
- an empty thread spool
- scissors
- a plank of wood 60" x 2" x $^3/_8$"
- a strip of wood 40" x $^5/_8$" x $^3/_8$"
- two small wooden blocks $^5/_8$" x $^3/_8$" x $^3/_8$" (one with a hole drilled through the center)
- a piece of wood 8" x 1" x $^3/_8$" (with two holes drilled $5^1/_2$" apart, as shown below)

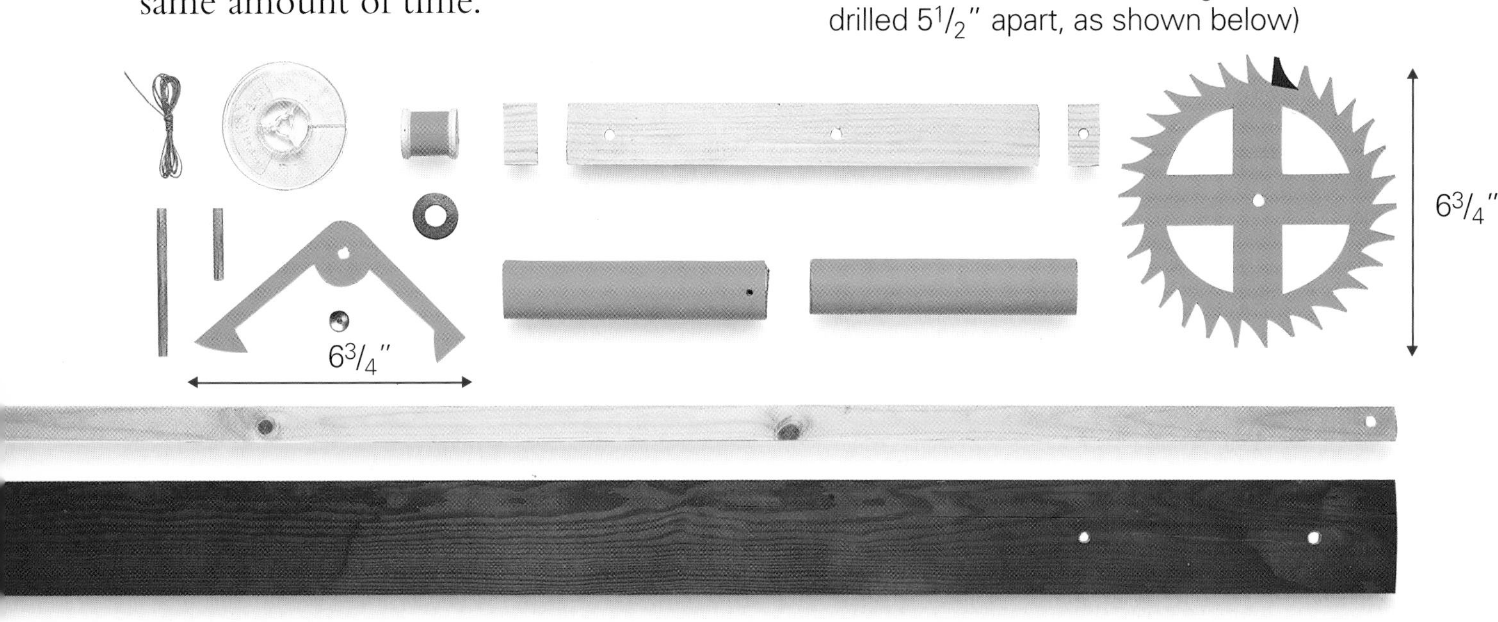

MAKE it WORK!

Sundials and water clocks are not nearly as accurate timekeepers as pendulum clocks. Pendulum clocks rely on a clever invention called an escapement, which lets the swing of a pendulum control how fast the hands turn. This model shows how an escapement works.

This is a difficult project, so you will probably need an adult to help.

1 The two main parts of the escapement are a toothed wheel and a rocking piece with two arms called an "anchor." Ask an adult to help you cut them accurately from stiff cardboard.

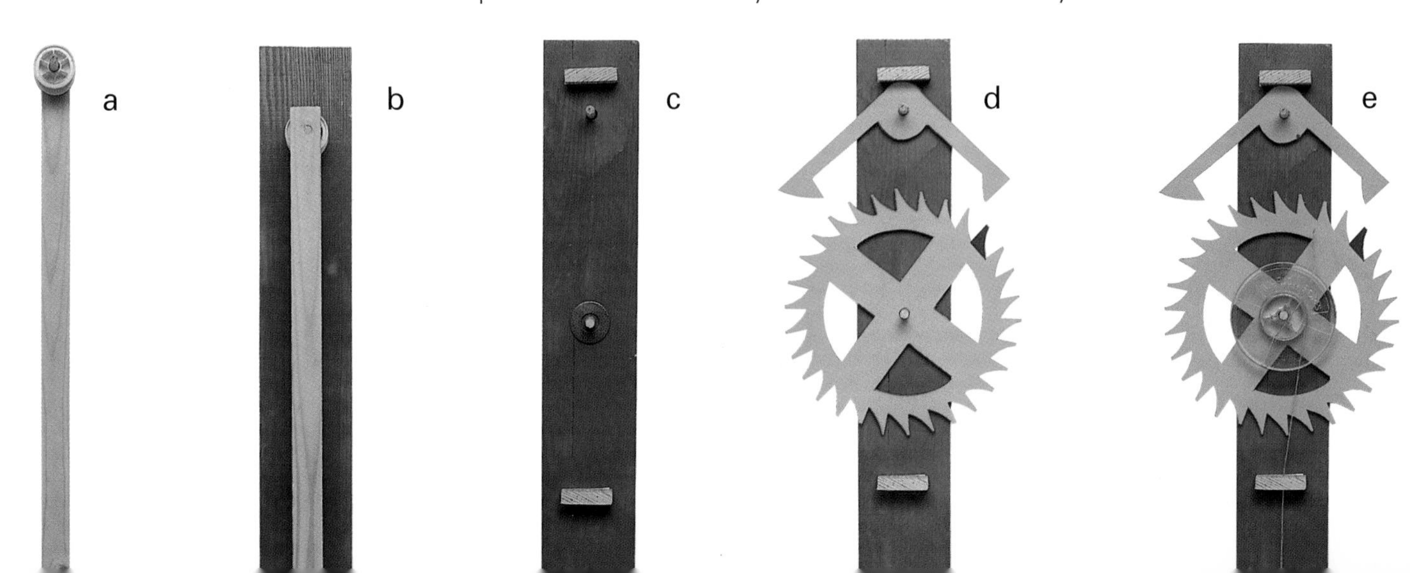

◀ As the pendulum swings, the anchor rocks to and fro. This releases the wheel to turn by one tooth at a time.

2 Drill a hole at the top of the strip of wood and another 2″ from the top of the plank. Glue a dowel in the hole in the strip, pass it through a thread spool (**a**) and through the hole in the plank so that the strip can swing freely (**b**). Glue a cardboard tube to the free end of the strip.

3 Glue the wood blocks $7^1/_2$″ apart on the other side of the plank so that the block without the hole is above the dowel.

4 Drill a hole for a dowel $5^1/_2$″ below the first. Glue in the dowel and place a washer on it (**c**).

5 Glue the anchor to the upper dowel so that it swings with the pendulum, and push the wheel onto the lower dowel so that it turns freely (**d**).

6 Glue the fishing line reel to the wheel and wind string onto it (**e**). Thread the string through the hole in the wood block. Tie the second tube to the end of the string.

7 Hold all the parts of the escapement together by slotting the dowels through the holes on the piece of wood. Pin the wood in place.

This pendulum clock is powered by the falling weight. You must wind the weight up to start the clock. The swinging pendulum releases the escapement at regular intervals, letting the weight drop a little each time. In a real clock the toothed wheel would turn **gears** to make the hands go around the face.

◀ Use bricks to support your clock. Try timing the beats of your clock with a stopwatch.

30 Electronic Clocks

Until about 300 years ago, the length of an hour was not fixed. It was often taken as a twelfth of the time between sunrise and sunset. This meant that hours were longer in summer than in winter. Then reliable **mechanical** clocks were invented. The hour scale on a clock dial has 12 equal divisions. The hour hand moves around the dial twice a day, dividing the day into 24 equal hours.

You will need

stencils
cardboard
a protractor
a craft knife
paints and pens
a battery for the clock mechanism
a clock mechanism with hands and a nut

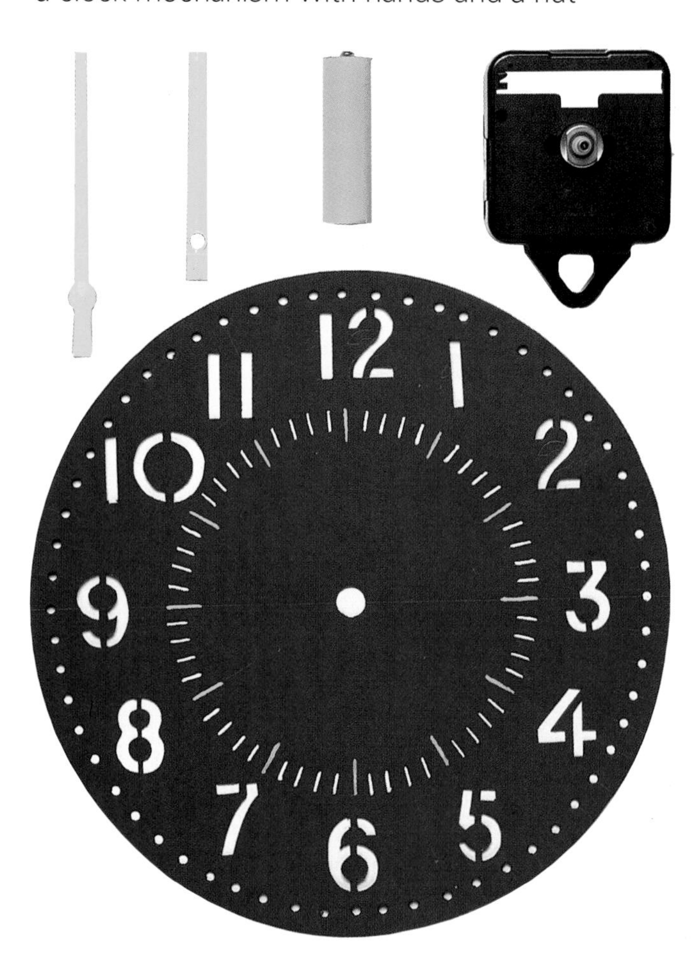

MAKE it WORK!

The first mechanical clocks were often placed in church towers so people could see them and hear their chimes. Falling weights powered the mechanism, and a swinging pendulum kept the hands turning at a steady rate. Modern clocks and watches are usually driven by a battery-powered electric motor. A tiny electronic **quartz crystal** is used in place of a pendulum. This **vibrates** at a steady rate, making sure the hands turn at the right speed. You can buy a quartz clock mechanism from a hobby shop. Why not design your own dial to go with it?

1 Draw around a large plate and cut out a circle of cardboard.

2 Use a protractor to pencil in the hour and minute marks around your clock face. The 12-hour marks are spaced 30° apart. The 60-minute marks are at 6° intervals around the circle.

3 Remove the nut from the central spindle of the clock mechanism, and make a hole in the center of the cardboard circle so that it will fit over the spindle.

4 Now you need to decide on the final design for your clock face. You could number each hour like the blue dial on this page (the numbers have been marked with a stencil and then cut out so they are see-through). Or you could just mark the hours and minutes with bold lines along the outer edge of the disk, like the orange dial shown on the left-hand page. Look at various clocks and watches for ideas.

5 Once you have designed and labeled your clock face, you are ready to fix it in place. Push the spindle of the clock mechanism through the center hole of the clock face from the back of the cardboard. Tighten the nut down on the front.

6 Push the hands onto the spindle and set them to the correct time. Start your clock by inserting the battery.

Electric clocks were first made about 40 years ago. Before then, most mechanical clocks were powered by weights or wind-up springs. Portable clocks, such as this alarm clock, had a balance wheel to operate the escapement mechanism. A balance wheel does the same job as a pendulum, by turning to and fro at a constant rate.

Pendulum clocks will not work at sea. In 1772, John Harrison won a £20,000 prize for inventing an accurate clock that could be taken on board a ship. This was very important because it allowed sailors to find their position accurately by working out their longitude.

How were clocks used to find longitude?
First you must set your clock to midday when the sun is overhead on the prime meridian (Greenwich Mean Time). Then you note the time by this clock when the sun is overhead at your location. Earth turns 15° an hour, so if the clock shows 1 P.M. you are 15° west of Greenwich. If it shows 9 A.M. you are 45° east.

32 Recording Time

Once a moment in time has passed it is gone forever. We cannot go back to it nor save it up to use again. But we can keep a record of what happened at that moment. This is what we do when we keep a diary or use cameras to take photographs and record videos.

MAKE it WORK!
We can "relive" special moments of the past by watching films and videos we have recorded. Videos play back at real time. This means that if it takes 10 seconds for an athlete to run a race, it will also take 10 seconds when a tape recording of the race is played back. You can, however, use a video machine to "reverse time." To record time you will need a video camera and some friends.

Shooting a sequence
Ask an adult to show you how to work the camera. Then record a short sequence of your friends throwing a ball into a bucket. Does it look the same forward as backward?

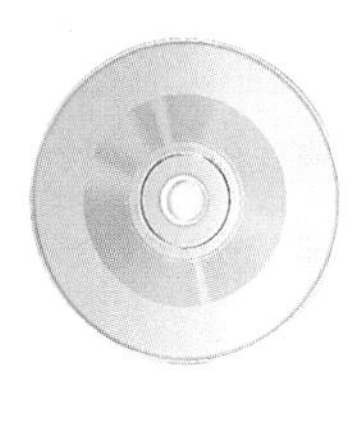

◀ Time can be recorded and played back using all sorts of equipment. Records and audio cassettes play back sound only. Videos and compact discs can record both sounds and pictures.

forward time

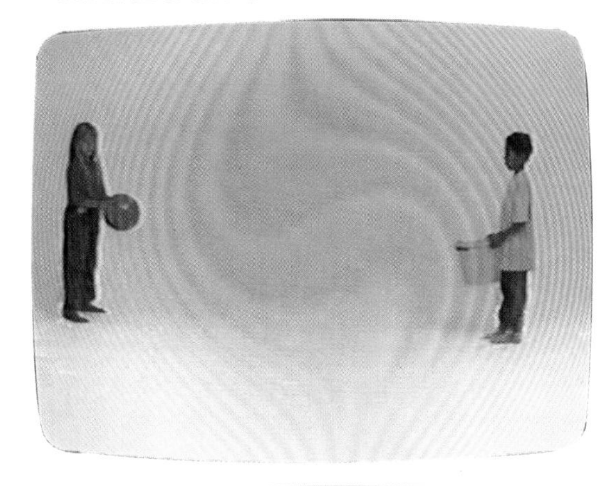

backward time

Video cameras take 25 or 30 pictures each second. Each picture records the scene at an instant in time. When you play back the tape, you see the pictures in the same order and at the same rate as they were recorded.

Playing with time

Now you can start to play tricks with time. Make time run backward, speed time up, or slow it down, by using the controls on the video recorder. You are not really making time do these things. You are creating an illusion on the screen. You can speed up time by showing more frames each second. If you slow the tape down far enough, you can see the changes between frames, and everything looks jerky.

The movie camera was invented just 100 years ago. When the first films were shown, people could not believe the images they saw. They would look behind the screen to see if anything was there! Films and videos are wonderful ways of recording events in time. Imagine how exciting it would be to have real films of the dinosaurs, Roman legions, or Columbus. Perhaps in the future your videos will be shown as a record of life in the 1990s.

34 Timetables

There are 24 hours in a day and seven days in a week. So each week you have 168 hours to fill. You spend about 60 or 70 of those hours asleep. Your waking hours are taken up by eating, studying, washing, pursuing hobbies, and seeing friends. But do you know how much time you spend on each of these activities?

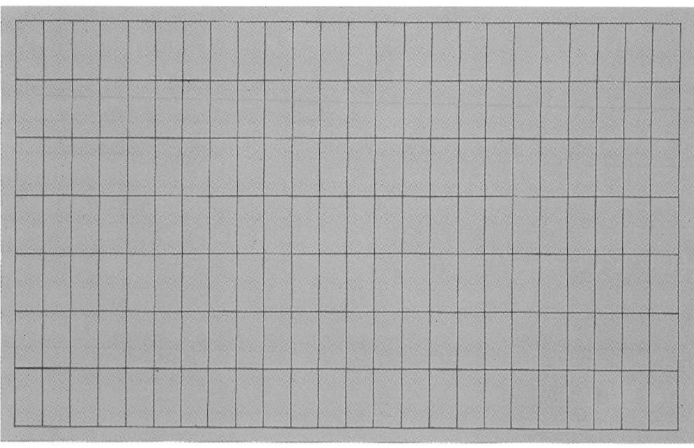

MAKE it WORK!
Do you spend more time watching television than playing sports? How much of your time is taken up by washing and eating? Try recording your timetable for a week to see where your time goes.

You will need

- scissors
- colored paper
- a large rectangle of cardboard
- glue
- a ruler
- pencils

1 Use a ruler and a pencil to draw the timetable grid on a sheet of cardboard. The grid has seven rows for the days of the week and 24 columns for the hours in each day. Number the rows and columns as shown.

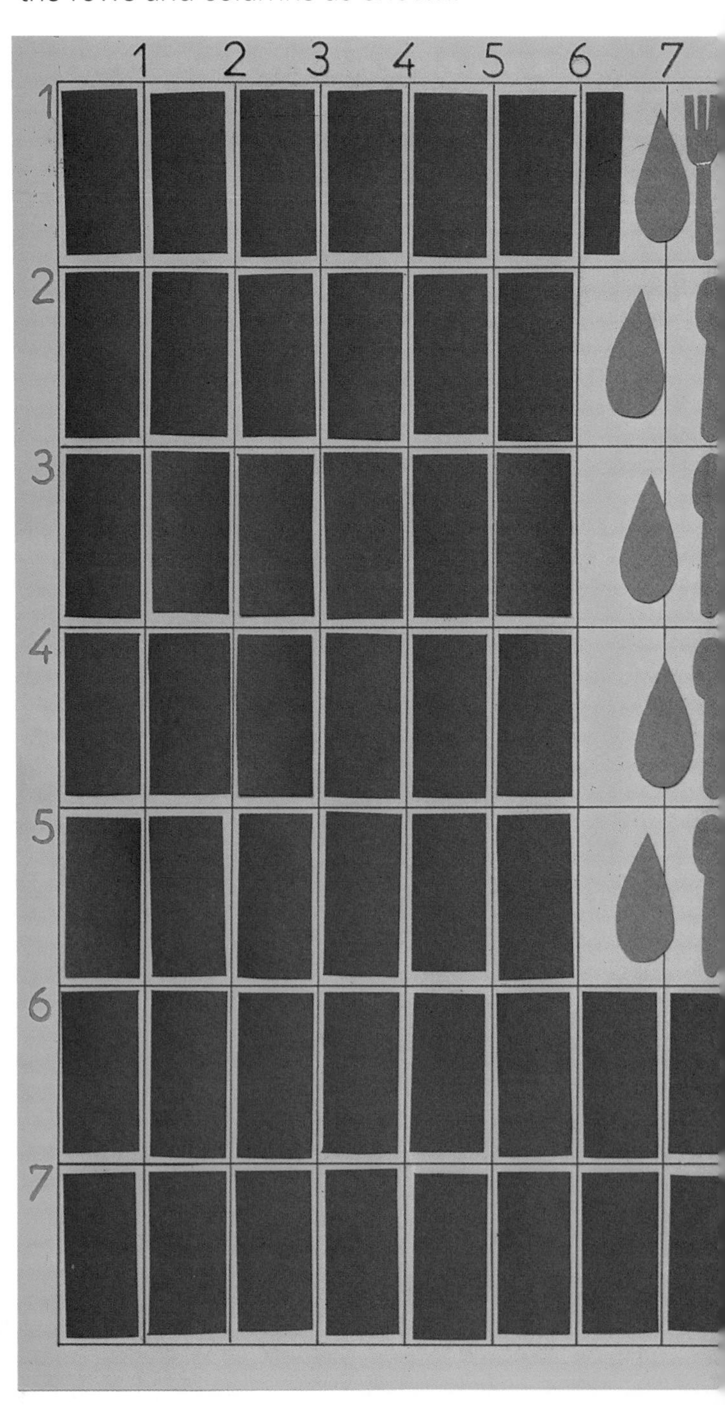

2 Make a list of the activities you want to record on the timetable. These are the activities you regularly spend time doing, such as playing tennis or football, watching television, attending school, or sleeping.

3 Design a "pictogram," like the ones shown on this page, for each activity. The blue rectangle stands for sleep, the tree for playing outside, the building for school, and so on. Make the pictograms from colored paper.

4 Each day, when you get the chance, place pictograms on the timetable to show what you have been doing. Use cardboard arrows to indicate when an activity has extended over several hours.

Did you know that during your life you will probably spend 25 years in bed, 10 years watching TV, 3 years eating, and 6 months in the shower?

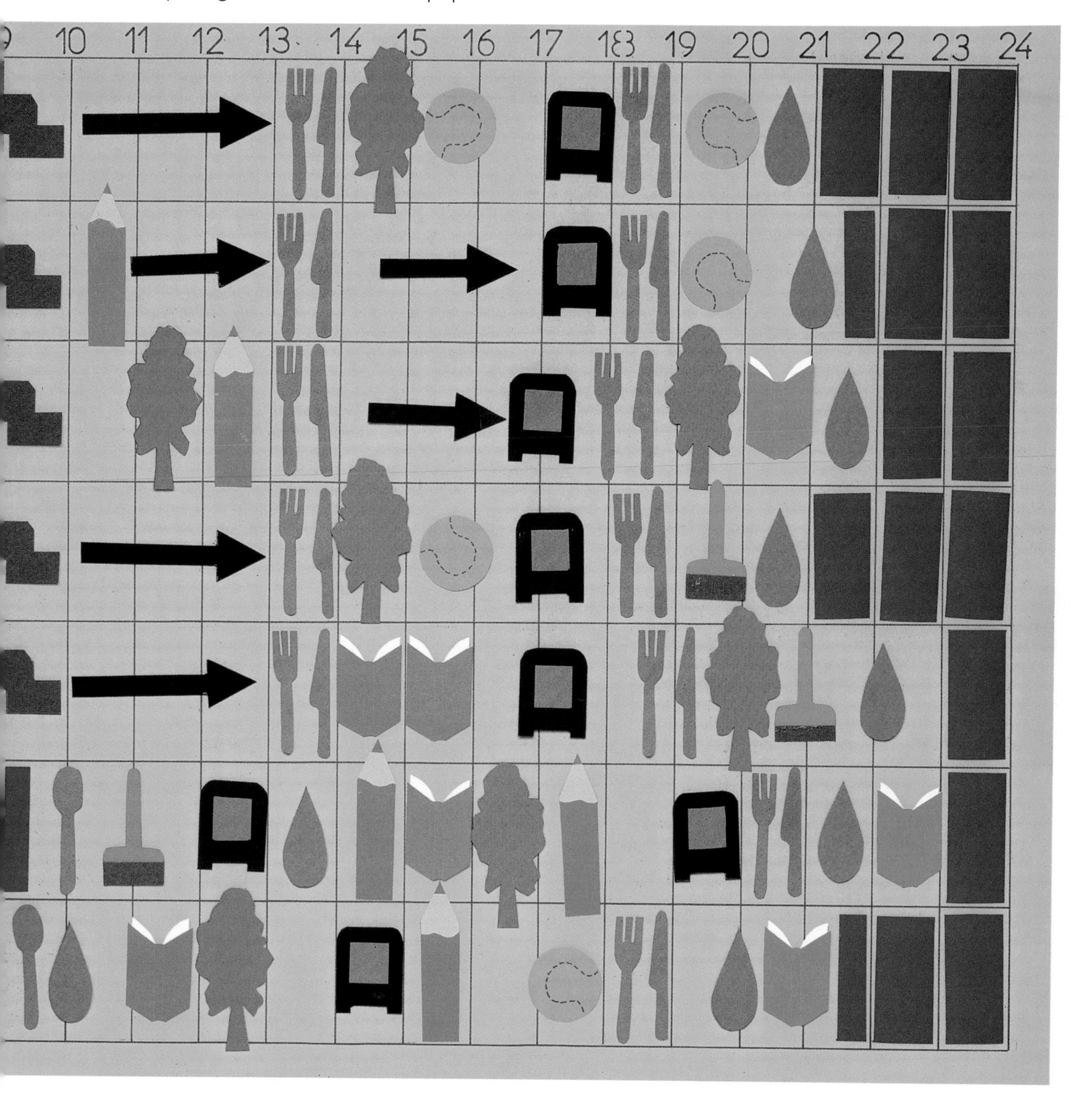

36 Time Capsules

Wouldn't it be amazing to travel back through time to the world as it was when our parents were young? We could see the clothes they wore, dance to the hit records they bought, and be present when great events that we now think of as history were actually happening. Sadly, this kind of time travel is impossible. But with the help of old films, newspapers, and other **memorabilia**, we can travel into the past in our imaginations.

You will need

a waterproof box with a lid
tape and plastic bags to seal things inside
items that give a picture of what your life is like in the 1990s—some ideas are listed below:

- A newspaper or comic from the week in which you sealed the box.

- A cassette tape on which you talk about yourself, your family, and your friends.

- Photographs of yourself, your home, your friends and pets. You could also cut pictures of famous people and places out of magazines.

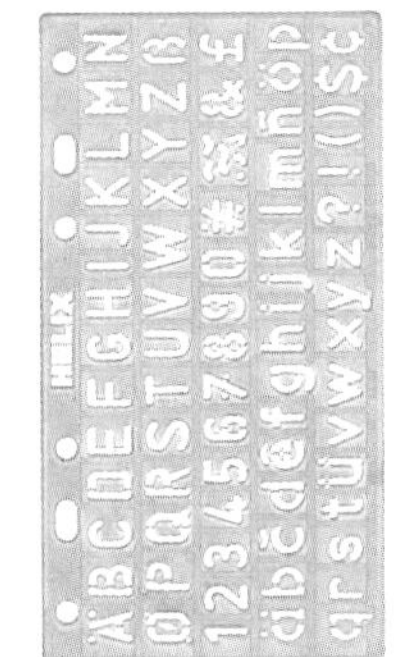

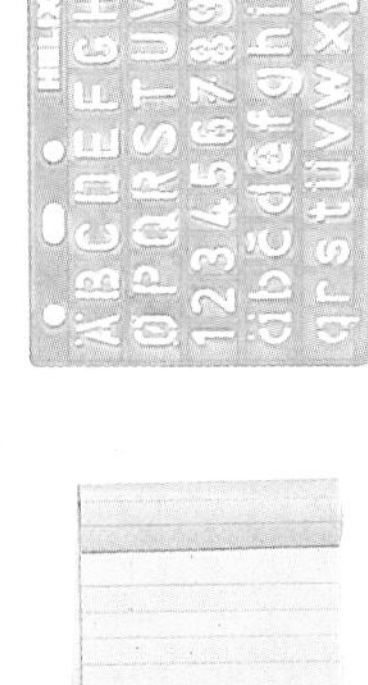

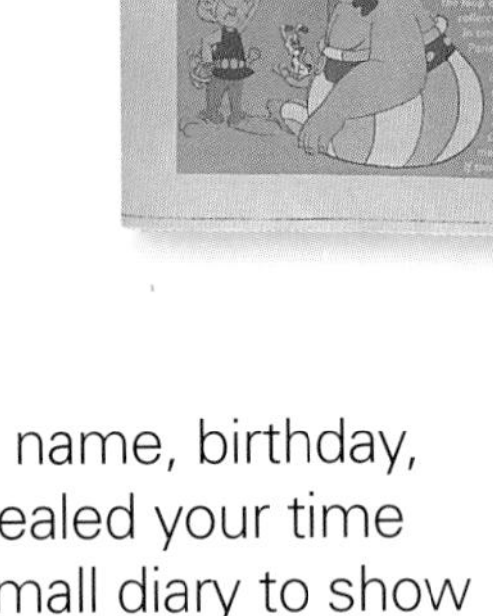

MAKE it WORK!

When **archaeologists** discover old relics that have been buried in the ground for hundreds of years, they are able to put together an idea of what life was like in times gone by. Your parents may still have some of their old records, clothes, and photos. These things give a fascinating picture of what life was like only 20 or 30 years ago. Why not prepare a time capsule of things from today that you or your family will find interesting in 10 or 20 years time?

- Notes recording your age, name, birthday, and the date and time you sealed your time capsule. You could write a small diary to show what you do in an average week.

- A tape of your favorite TV shows and home videos of you and your family.

Seal everything into plastic bags and pack the box carefully. Seal the lid in place with waterproof tape.

Now you must decide where you want to hide your time capsule so that it will travel safely into the future. You could bury it in the yard, hide it away in the attic, or leave it with a teacher at school with instructions that it should not be opened for 20 years.

Whoever finds your capsule in the future will be fascinated by the story that your objects and messages tell.

When old buildings such as churches are repaired, time capsules hundreds of years old are sometimes found hidden in the walls and rafters. Of course, these do not contain videos or tapes, but there is often a letter giving the date when the building was finished and listing the names of the builders.

38 Geological Time

How old is Earth? Until the last century most people believed that Earth was just 6,000 years old. But now scientific study of rocks has shown that it is much, much older. Earth formed from clouds of dust and debris in space 4.5 billion years ago!

You will need

scissors
lengths of thin dowel
paints and paintbrushes
a 6″ length of $^{3}/_{8}$″ dowel
a piece of wood $12^{3}/_{4}$″ x $^{3}/_{4}$″ x $^{3}/_{8}$″
two pieces of wood 6″ x $^{3}/_{4}$″ x $^{3}/_{8}$″
glue
cardboard

1 On cardboard, draw a 16″ diameter circle. Draw pencil lines from the centre to the edge every 5°.

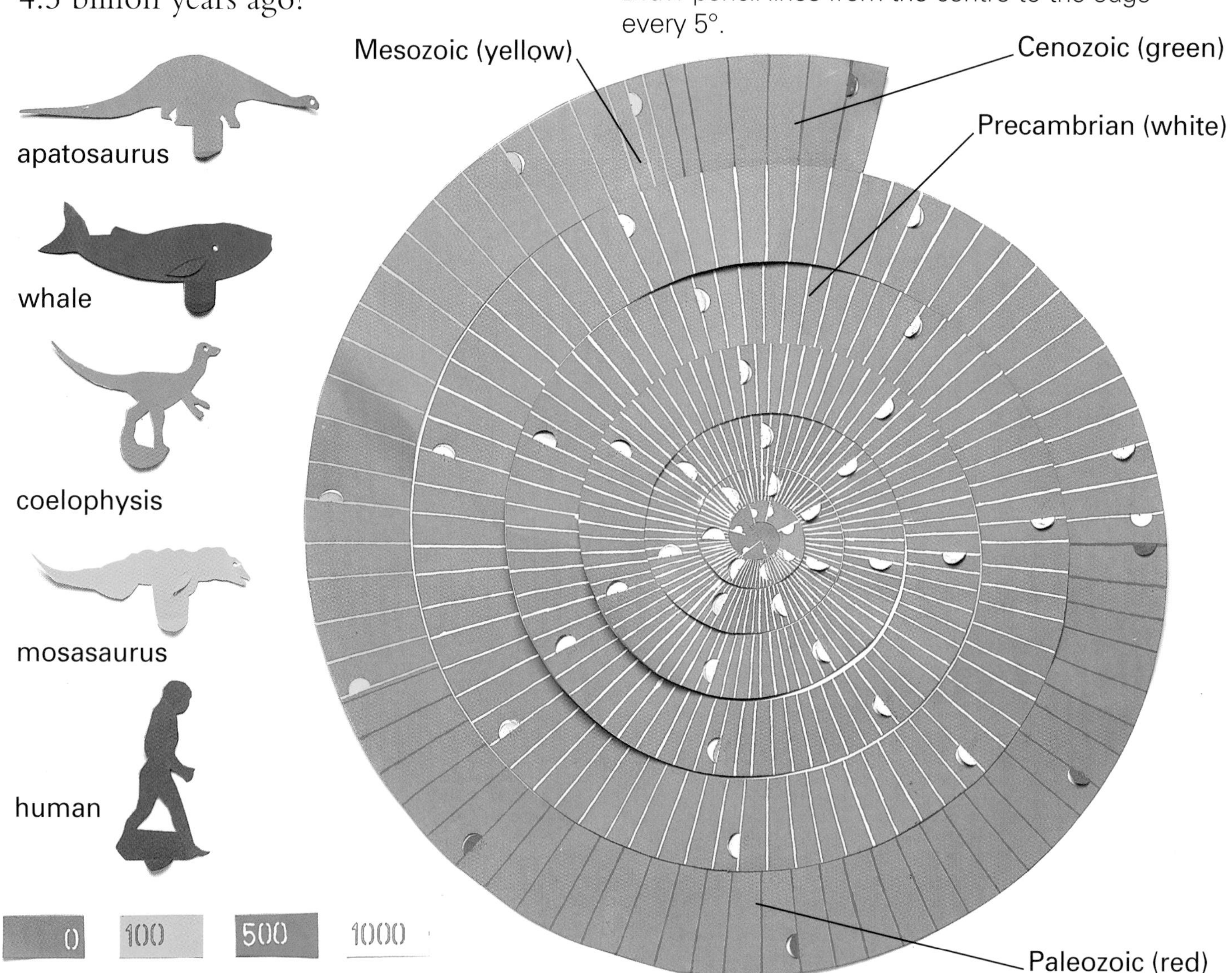

MAKE it WORK!

Looking back in time is like looking down a spiral time tunnel. The recent past seems quite clear, but the farther back you look the more the years seem to merge together. The huge intervals of **geological** time are measured, not in years, but in periods of billions of years.

2 Draw a spiral with about seven turns. Paint the colored lines as shown. These lines are marked every 10 million years. The present day is at the wide end of the spiral. The lines get closer together as you spiral into the center, 4.5 billion years ago. Now cut out the spiral.

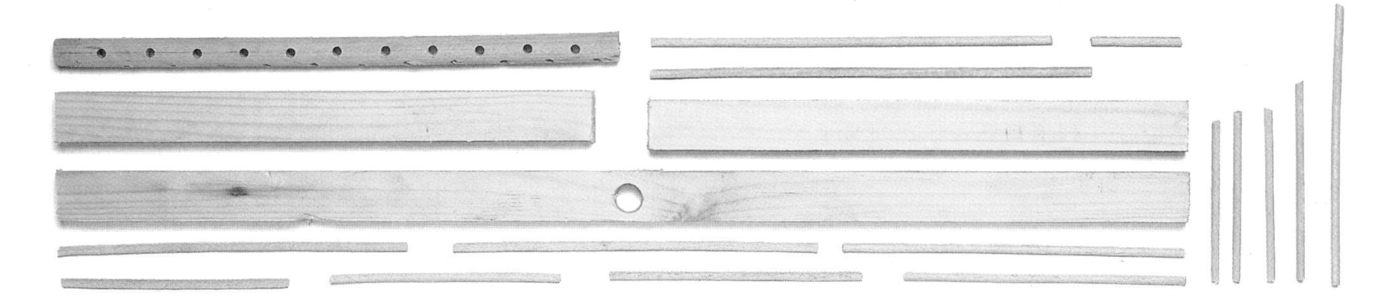

The small semicircular tags appear every billion years and also mark special time periods called eras. There are four great eras marked by white, red, yellow, and green lines on the spiral:

Precambrian	4,600—570 million years ago
Paleozoic	570—240 million years ago
Mesozoic	240—63 million years ago
Cenozoic	63 million years ago—present

3 Use the pieces of wood and dowel to make a stand for your time spiral, as shown below. Ask an adult to help you drill the holes in the wood. Fix the pieces together with glue. You can now glue the spiral in place.

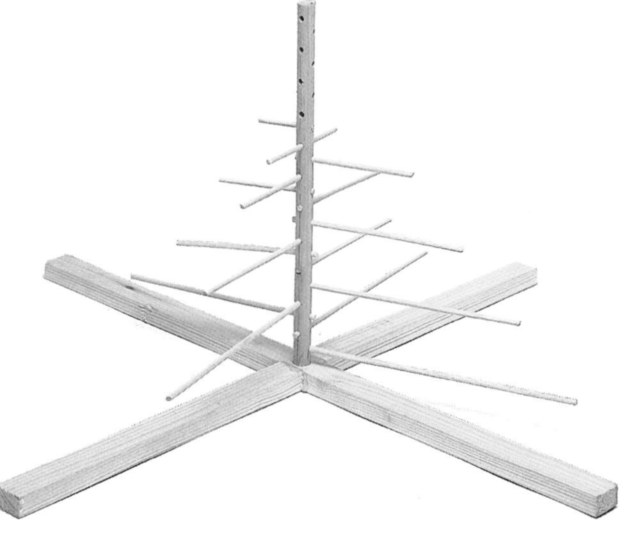

4 Draw and cut out the various animal species shown, and place them in their appropriate periods on the time spiral. Refer to books about **fossils** and the history of life on Earth to find other species to add to your model.

The first evidence for life on Earth has been found in rocks 3.5 billion years old. For most of Earth's history living things were very simple, similar to the bacteria and viruses that give us coughs and colds today. The first land plants appeared about 400 million years ago. The age of the dinosaurs lasted from about 240 to 63 million years ago. Since then we have been in the age of mammals, including the great ocean whales and ourselves. Modern humans appeared very recently in Earth's history, probably less than a quarter of a million years ago.

40 Order and Disorder

In time, everything changes. Many things decay, rust, or erode. If you see a picture of a shiny new bicycle next to a picture of the same bicycle looking rusty and unusable, you can probably guess which picture was taken first. But you could be wrong if effort had been put into renovating an old bike.

Creating order

Seeds create order as they **germinate** and grow. You can observe this process in action by germinating some cress.

You will need

cotton
a shallow bowl or dish
a packet of cress seeds

1 Line the dish with cotton and moisten the cotton with water.

2 Sprinkle a layer of cress seeds onto the moist cotton.

3 Leave the dish in a warm place and wait for the seeds to germinate.

You will need to water your cress every day. It should take a few days for your seeds to germinate and about a week for them to grow to the height shown in the middle of the opposite page.

MAKE it WORK!

Everything is made up from very small particles called molecules. Processes such as burning and decay break down complex molecules into simpler ones. For example, when gasoline is burned, it is broken down into carbon dioxide and water. On the other hand, in living things, simple molecules can change into complex structures.

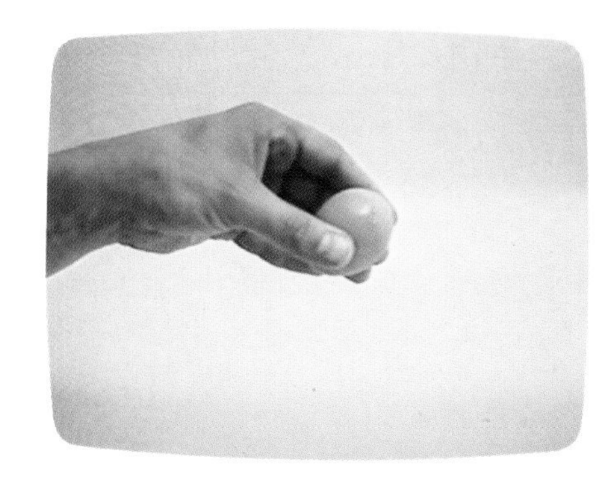

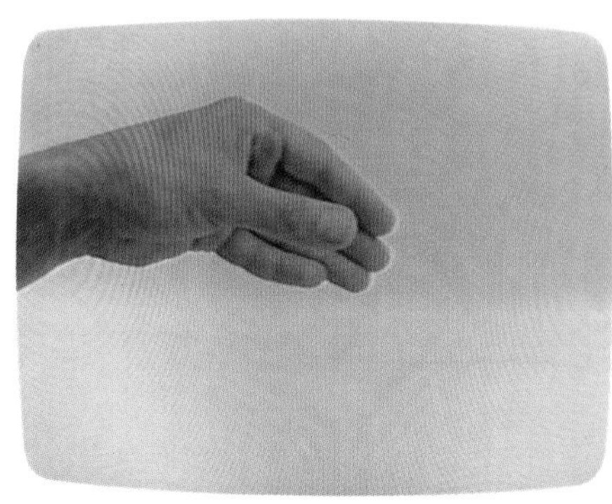

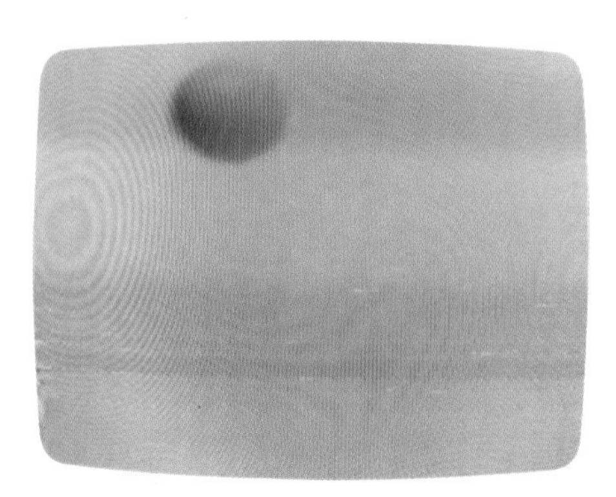

The seeds sprout and tiny green leaves unfold. Leaves are the new plant's food factories. They trap the energy of sunlight to make food from carbon dioxide in the air and water drawn up through the roots.

Plants and other living things are called organisms. During their lives they organize non-living materials to make living materials. But if you don't water your cress, the seedlings soon turn brown and die. Once the life process has stopped, order decays into disorder.

What is disorder?

The scientific word for disorder is entropy. The more disordered something is the greater its entropy. To create greater order you need to concentrate **energy**. Animals feed on plants and use their chemicals to grow their own bodies. When an animal dies, its molecules are spread out and disordered. Scientists believe that entropy (disorder) is increasing, so the universe is said to be running down.

▼ without water the cress dies

Creating disorder

Have you ever smashed an egg? Once it has splattered on the floor, the only way to put it together is to make time run backward. You can do that with trick photography, but not in real life. Smashing an egg is a one-way process.

If we look at a series of pictures of the event, like those shown along the bottom of these pages, we have no difficulty in deciding on the direction of time. The whole, ordered egg comes before the smashed, disordered egg.

Scientists sometimes think of time as the fourth dimension. For example, to describe the position of an aircraft we need four pieces of information: its latitude, longitude, height (three dimensions of space), and the time it is at that position (the fourth dimension). The fourth dimension is very different from the other three. We can choose whether to go up or down, right or left, forward or backward in space. But we can only travel forward through time at the same speed as everyone else.

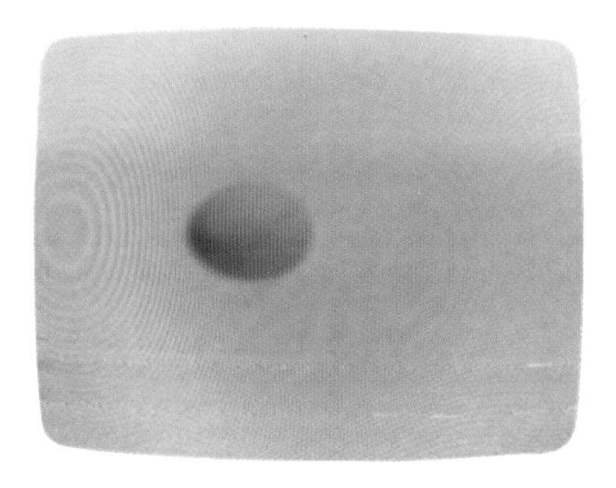

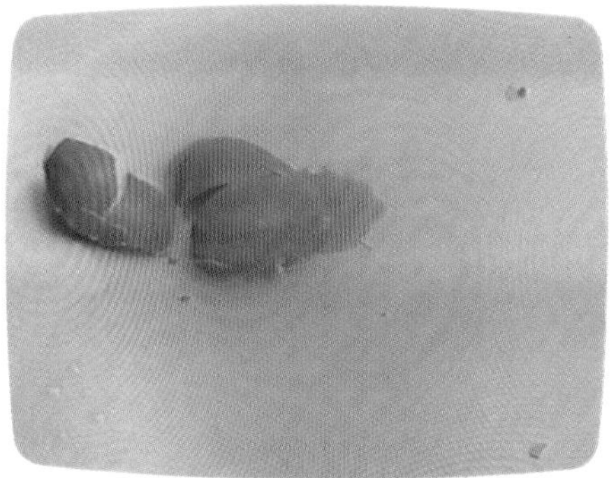

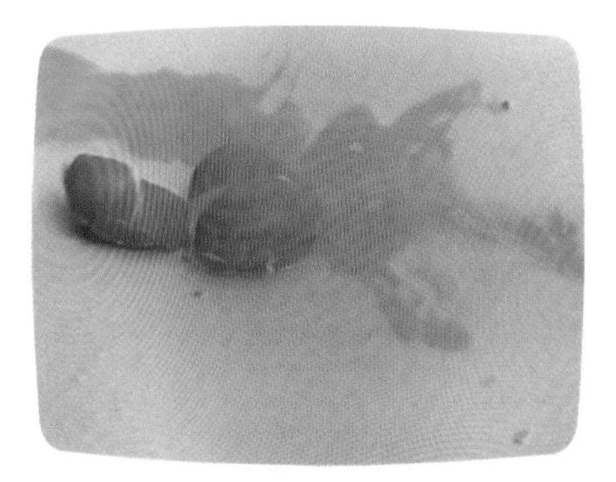

42 Light from the Stars

Imagine flashing a message to the stars with a laser. How long would you have to wait to receive a reply? Apart from the sun, the star nearest to Earth is four **light-years** away. So, if someone is out there waiting for your message, it would take at least eight light-years for your greetings to be returned!

MAKE it WORK!

When you look at the stars you look back in time. The farther you look out into space the farther back into the past you see.

You will need

- marbles
- thin string
- a piece of plywood
- dowels of different lengths
- a sponge ball to represent Earth
- glue
- a pin
- a drill
- paints
- beads

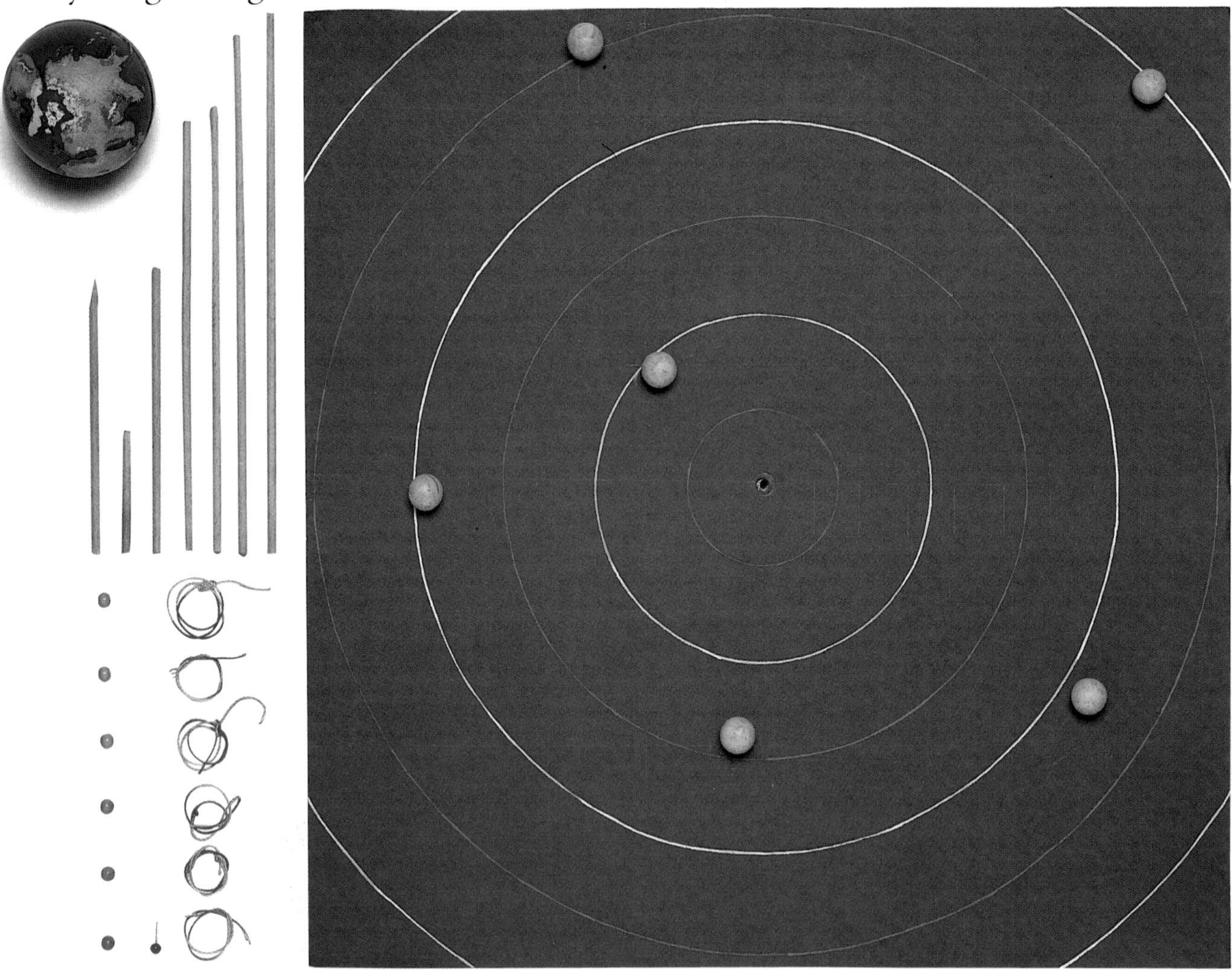

Light is the fastest thing in the universe, but the distances to the stars are so great that it takes years for their light to reach us. We see the stars, not as they are now, but as they were years ago when the light left them.

1 Paint the plywood dark blue or black. Mark it with six circles, as above. Use a pin and a length of string to help you draw neat circles.

2 Drill a hole in the center of the plywood and six more at different distances from the center. The holes should be big enough to fit the dowels.

3 Glue dowels into the holes. Glue Earth onto the top of the central dowel and the stars (marbles) to the other dowels.

4 Push a pin into your location on Earth. Tie a string between each star and the pin. Thread a bead onto each string before you finally fix it in place.

The strings show the straight lines along which light travels from the stars to your eyes. Push the beads up to the stars. Move the beads equal distances to show light traveling at the same speed from each of the stars. Light from the nearer stars will reach Earth before the light from the more distant ones.

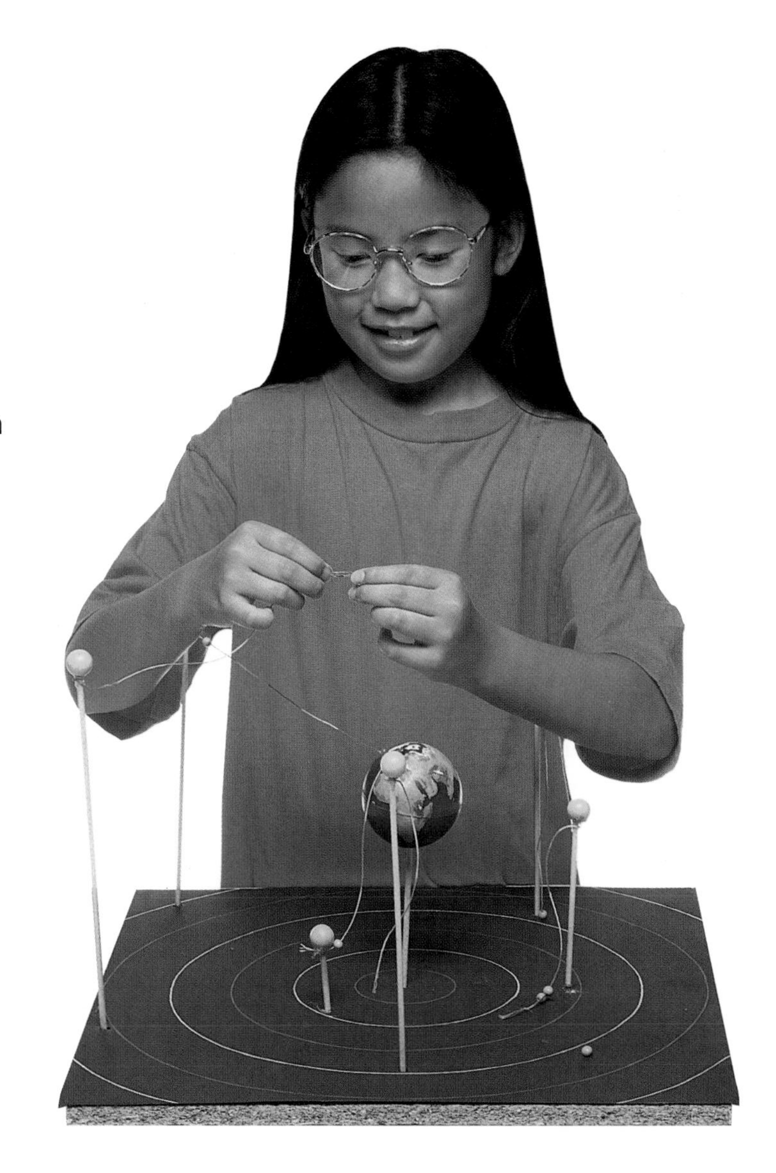

The sun is 93 million miles from Earth; that is eight light-minutes. If you are reading this book by sunlight, then the light now entering your eyes left the sun's surface just eight minutes ago.

Astronomers use powerful telescopes to look at the most distant objects in the universe. These objects are so far away that their light takes almost 15 billion years to reach Earth. Using radio telescopes astronomers can detect the energy from the ***big bang*** *that started the universe 10 billion to 20 billion years ago.*

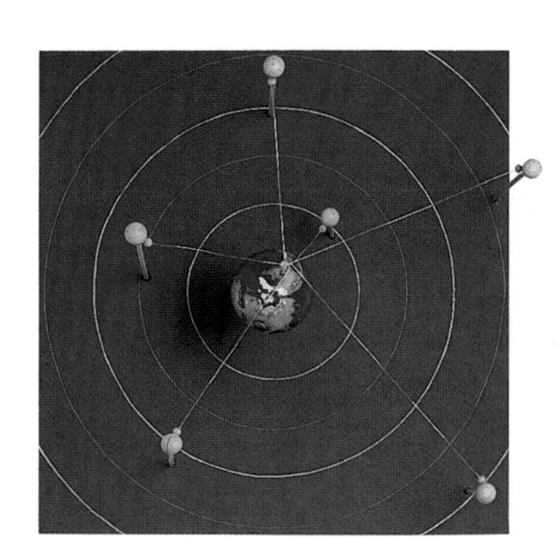

▲ Stars are huge balls of gas that give off light energy in all directions.

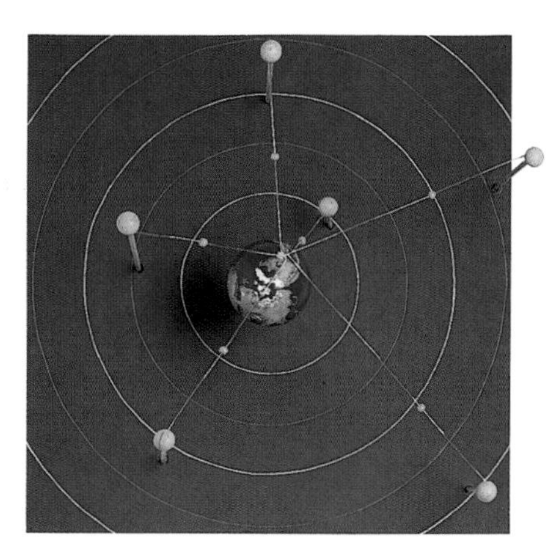

▲ The light we see travels toward Earth at about 186,000 mi per second.

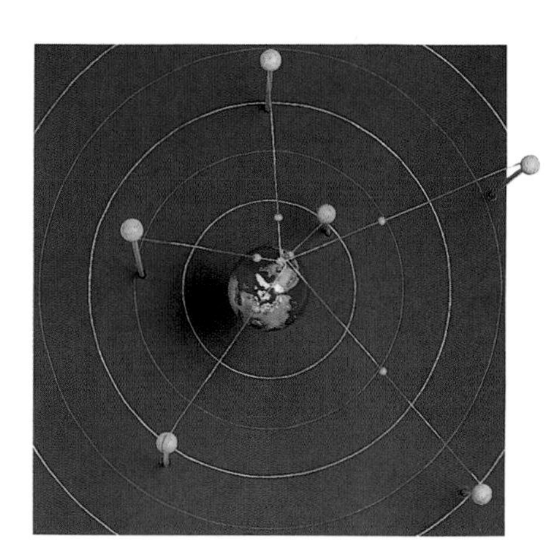

▲ The light from the nearest star reaches Earth first.

44 Time Twins

Time flies when you are enjoying yourself. An hour spent playing your favorite computer game can seem like five minutes. But sometimes time drags—five minutes in the dentist's chair can seem like an hour!

MAKE it WORK!

Bill and Ben are identical twins born on the same day. In an imaginary world Ben's time passes twice as fast as Bill's time. When Ben is starting school, Bill is still in diapers. When Ben is driving a car, Bill is still on roller skates. When Ben is a middle-aged man, Bill is going out on his first date. This flick book shows how the appearances of Bill and Ben change. Starting as identical babies they age at different rates.

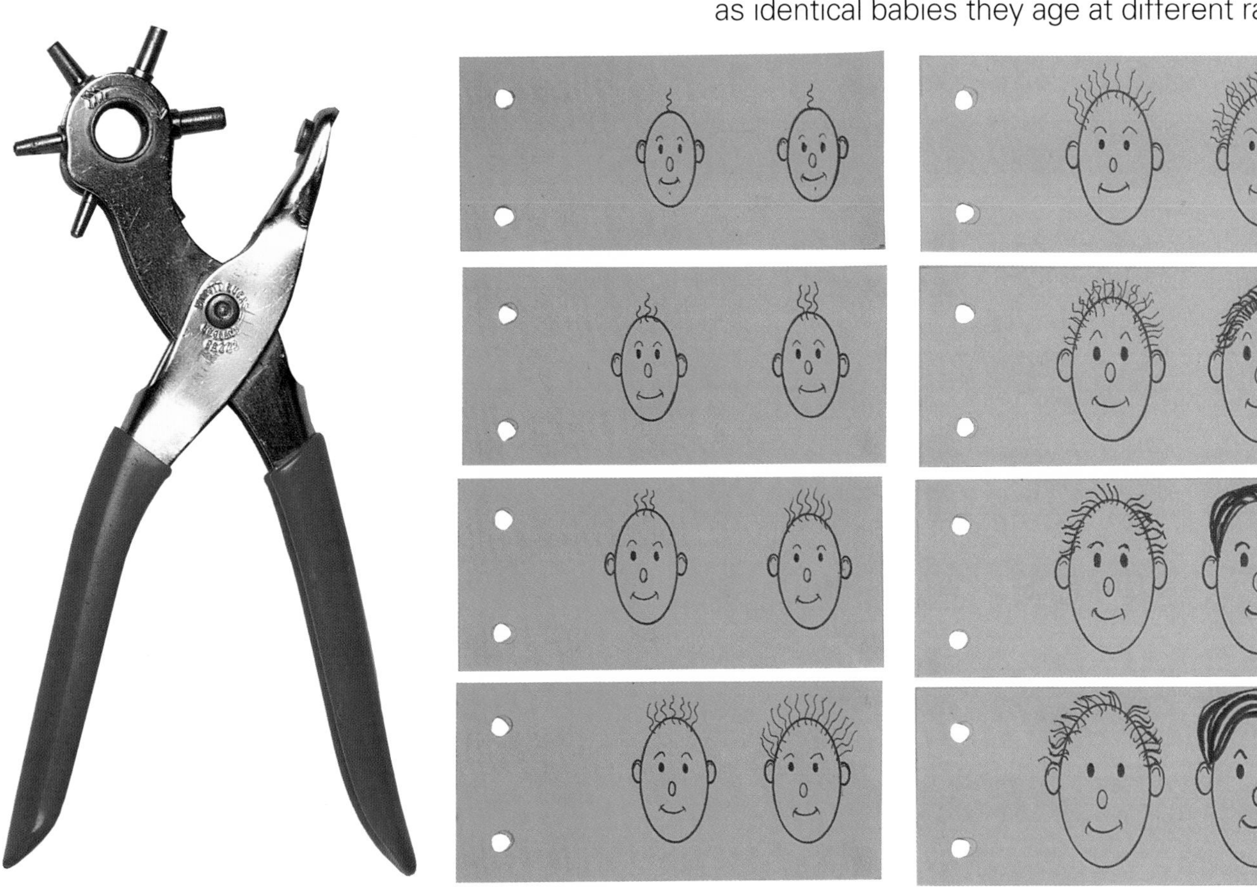

Of course every hour is exactly the same length. Your watch does not slow down when you are doing something you hate, or speed up when you are enjoying yourself. Imagine what the world would be like if people's clocks ran at different speeds. People whose clocks ran fast would live their lives more quickly than people whose clocks ran slow. They would grow up and age at different rates.

You will need

a hole punch	pens	nuts
thin cardboard	scissors	bolts

1 Cut 18 thin cardboard rectangles approximately 4″ x $1\frac{1}{2}$″.

2 Draw the twins Bill and Ben on the cards, as shown. They start as identical babies but age at different rates.

3 Punch two holes in each card. Fix them together with nuts and bolts to make a book.

Scientists have proved that Einstein's theory works. They sent an accurate clock on a journey around the world in a fast aircraft. When the aircraft landed, the clock on board was a fraction of a second slower than a clock that had stayed on the ground.

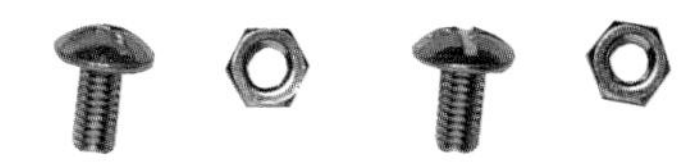

4 Flick through the book to see how the twins age.

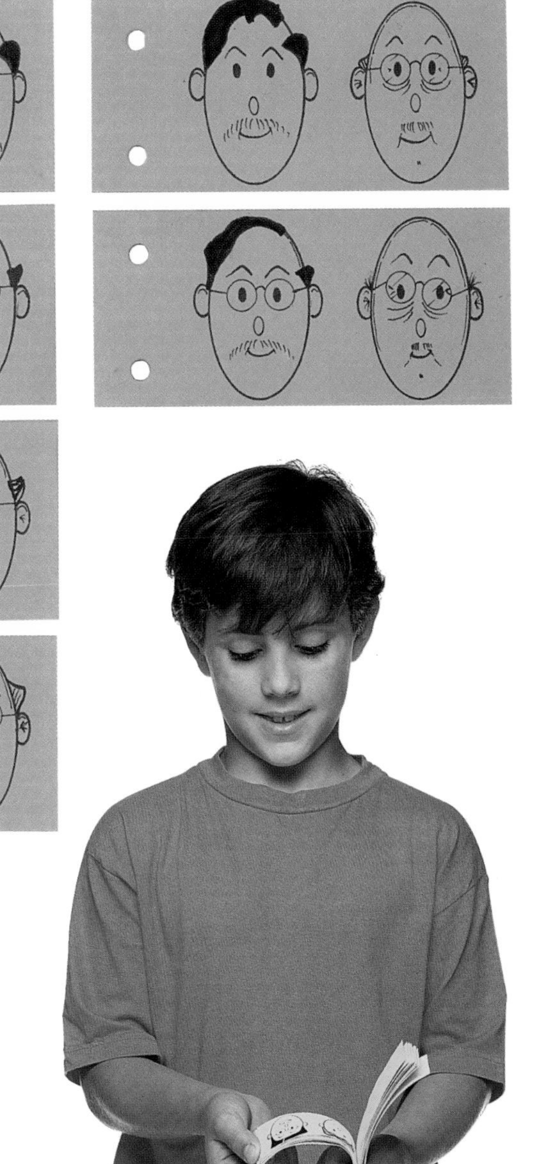

Science fiction or science fact?

The story of Bill and Ben sounds like science fiction. But, in **theory** at least, it might one day become a science fact. At the beginning of this century the great physicist Albert Einstein worked out the theory of relativity. This suggests that time need not pass at the same rate for everyone in the universe. If you made a journey away from Earth in a spaceship that could travel nearly as fast as light, when you returned you would have aged less than people who stayed behind. You might even be younger than your own children!

Glossary

Archaeologist A person who digs up and studies ancient remains.

Axis A line that something turns around. Earth's axis is an imaginary line running between the South Pole and the North Pole. Earth makes one full turn around this axis once a day.

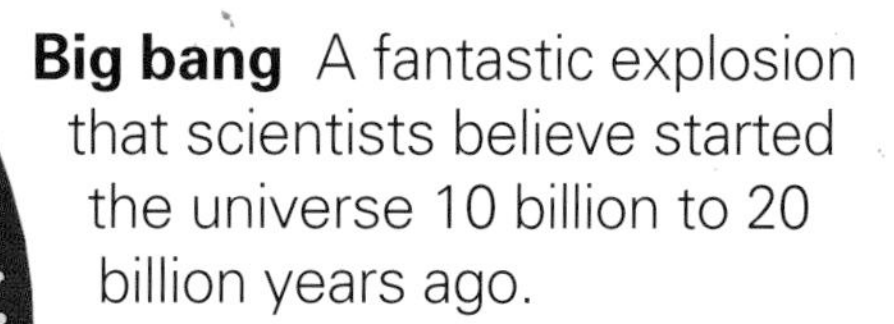

Big bang A fantastic explosion that scientists believe started the universe 10 billion to 20 billion years ago.

Calendar A chart on which the days, weeks, and months of the year are set out. Many calendars show one month per page.

Calibrate To make or check the marks on a measuring scale, such as a ruler or a sundial.

Celestial This means anything to do with the sky. We can imagine that the stars are fixed to a celestial sphere that turns around Earth. Actually the stars are at different distances from Earth, and it is Earth that turns. The stars appear to move around the celestial pole.

Century A period of 100 years. The twentieth century began on January 1, 1901, and will end on December 31, 2000.

Compass A magnetic compass has a needle that points toward the North Pole. It is used for direction finding.

Counterclockwise When something turns counter-clockwise it turns in the opposite direction to the hands on a clock.

Day The time it takes for Earth to spin around once on its axis; 24 hours.

Diameter A line that passes through the center of a circle, cutting it in half.

Digital A digital watch displays the time with digits instead of a clock face with moving hands. For example, half past eight would be displayed as 8:30.

Energy When something has energy it can make things move or change. Energy is needed to turn the hands of a clock. This energy can be provided by a falling weight, a wound-up spring, or an electric battery.

Equator Earth's equator is a line halfway between the poles. It divides the Northern Hemisphere from the Southern Hemisphere.

Fossil The remains of a once-living thing that has been preserved in rock.

Gear Gears are toothed wheels that link together to carry turning movement from one place to another. Inside a clock, gears are used to change the speed of movement to make the hour, minute, and second hands turn at different rates.

Geological This means anything to do with Earth's rock. It takes millions of years for new rocks to form and old ones to change. The huge time periods during which these events happen are known as geological time.

Germinate To start to develop or grow.

Hemisphere Half a sphere, like a dome. Earth is divided into the Northern Hemisphere and the Southern Hemisphere by an imaginary line, the equator.

Hour One twenty-fourth of a day.

Latitude A measure, in degrees, of a position on Earth, north or south.

Leap year An extra day, February 29, is added to the calendar in a leap year, which comes every four years. This keeps the calendar in step with the seasons. Earth takes just under $365\frac{1}{4}$ days to orbit the sun. The quarter-days add up to an extra day every four years.

Light-year A measure of distance. One light-year is the distance light travels in one year.

Longitude A measure, in degrees, of your position around Earth, east or west.

Lunar To do with the moon.

Mechanical To do with machines. A mechanical clock is a machine for telling the time.

Memorabilia Objects from the past that remind us of how life was at that time.

Minute One-sixtieth of an hour.

Month A lunar month is the number of days it takes for the moon to orbit Earth. We divide our year into 12 calendar months of 28-31 days.

Orbit The curved path followed by a planet, a moon, or a satellite as it circles around a more massive body.

Planisphere A circular star chart designed for finding the stars at different times of night. The 12 months are marked around the edge of the chart. A second disk with an oval window sits on top of the chart. The time is marked around the edge of the second disk. When you line up the time with the date, the stars that are then visible appear in the window.

Quartz crystal A piece of transparent mineral found in rocks such as granite. When a quartz crystal is put in an electric circuit it can be made to vibrate at a steady rate. This vibration is used to make clocks and watches keep accurate time.

Radius The line between the center of a circle and its edge.

Second One-sixtieth of a minute.

Solar To do with the sun.

Sundial A sundial shows the time by the movement of shadow cast by the sun. The shadow is cast onto a scale by a pointer called a gnomon.

Theory An idea that tries to explain something. Scientific theories have to be proven by experiments before they are said to be true.

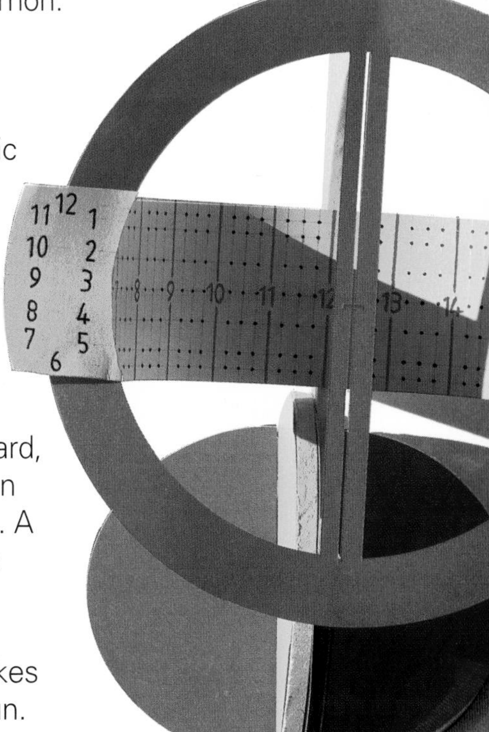

Vibrate To move backward and forward, repeating the motion over and over again. A swinging pendulum is vibrating.

Year The time it takes Earth to orbit the sun.

Index

NEW HANOVER COUNTY PUBLIC LIB.
3 4200 00466 5671